VṚKṢĀYURVEDA
(A Treatise on Plant Science)

Indian Medical Science Series No. 38

Vṛkṣāyurveda of Parāśara

(A Treatise on Plant Science)

Sanskrit Text, English Translation and Notes with Comparative References to Modern Botany

edited by
N.N. Sircar
and
Roma Sarkar

Sri Satguru Publications
A Division of
Indian Books Centre
Shakti Nagar, Delhi, INDIA

Published by
Sri Satguru Publications
A Division of
Indian Books Centre
24/4, Shakti Nagar, Delhi-110 007

INDIA

First Edition : Delhi, 1996

ISBN 81- 7030-441-5

" The Publishers is grateful to the Indian National Science Academy for Permission to Print this monograph completed under the auspices of Indian National Commission for History of Science. The Academy is not responsible for the views expressed in the book."

PRINTED IN INDIA

This monograph is dedicated to the sacred memory of J.N. Sircar Viṣagratna who was the main architect for the foundation of this work

Contents

Foreword

The text of the Vṛkṣāyurveda of Parāśara edited by N.N. Sircar and Roma Sarkar is a remarkable contribution in the field of Botany in ancient India. The work consists of the text of the Vṛkṣāyurveda in Sanskrit, an English translation and critical notes. A dissertation under the heading "An Introduction to the Vṛkṣāyurveda of Parāśara" by the late N.N. Sircar appeared in the Journal of the Asiatic Society of Bengal Vol. XVI. No. 1, 1950. In this article a preliminary account of the text-critical apparatus and of the contents of the Manuscript were furnished. It was stated therein that the original Ms was discovered by the late Vaidyaśāstri Jogendranath Sircar Viṣagratna, father of the late N.N. Sircar. The handwritten Devanagari copy of the Ms as now in possession of Roma Sarkar, contains five parts i.e. Bījotpattikāṇḍa, Vānaspatikāṇḍa, Vānaspatyakāṇḍa, Gulmakṣupakāṇḍa and Virudhavallīkāṇḍa. The text contains 65 folios in Devanagari script. The sixth and last part, the Cikitsitakāṇḍa as mentioned in the content (vs. 15 Ch. 1, Bījotpattikāṇḍa) is not available in the present Ms. The text ends abruptly with the Vīrudhavallīkāṇḍa. Examination shows that the present Ms unlike majority of the Mss does not contain an introduction regarding the author and of his lineage; a colophon at the end containing the date of composition is found missing too. But by paying homage to Brahmā, the Creator of the Universe, and by referring to the work as a subsidiary branch of the Atharva Veda it has in a way fulfilled the object of introduction. Each chapter of the Ms begins with the caption "ityāha Parāśara" and the concluding lines everywhere contain the statement "iti Parāśarakṛte Vṛkṣāyurvede".. This Ms therefore, establishes Parāśara's authorship of the Vṛkṣāyurveda.. In determining the authenticity of the text we have first of all to examine the validity of the tradition fathering the authorship of the Vṛkṣāyurveda..., to Parāśara, and secondly to examine the contents of the Ms to ascertain the nature of the data furnished by it.

The name of Parāśara has come down to us from the hoary past ages as an authority on politics, smṛti, astrology, astronomy, agriculture, meteorology, and the science of prognostications. Parāśara is also credited with works on medicine and on the science of the use of whey (takrakalpa)[1]. Parāśara has been referred to with reverence in the Ṛgveda[2], the Taittirīya Āraṇyaka[3], the Bṛhadāraṇyakopaniṣad[4], Mahābhārata,[5] Yājñavalkyasaṁhitā,[6] Carakasamhitā[7] and Bṛhatsāmhitā.[8] Bhaṭṭotpala in his commentery on the Bṛhatsamhitā Ch. 29 cites a verse from Parāśara dwelling on the features of plants (vs 14). Bhaṭṭotpala can be placed in 976 A.D. from his own account.[9]

Thus we find that there is a great tradition associated with the name of Parāśara who might be a progenitor of a school and it can at best be inferred that the author of the present text may belong to that school.

The provenance of the Ms was probably Coochbehar[10] in North Bengal where Vaidyaśāstri Jogendranath Sircar spent

---

1. The History of Indian Medicine. G. Mukherjee Vol. 1. p 44.
2. Rgveda VII. 18. 21.
3. Taittirīya Āraṇyaka I.I. 3. 37.
4. Bṛ. Ar. Up. II. 6. 2; IV. 6. 3.
5. Sabhāparvaṅ Ch. 113. vss 15-27; Chitrasala Edn
6. Yāj. I 5.
7. Sūtrasthāna, Ch. I.
8. Ch. 29.
9. Preface to the Bṛhatsamhitā. Sarasvatibhavan series 96.
10. Vide the Catalogue of Sanskrit Mss in the Sahityasabha of Coochbehared by D.K. Kanjilal and published by the Sahityasabha, Coochbehar 1977.

long span of his life. Coochbehar under the rule of the Maharajas from the late 13 the cent onwards grew up as a centre of study where scholars from distant parts of India converged. As a repository of the various texts on Nyāya, Vyākaraṇa, Kāvya, Alamkāra, Jyotiṣa, Mimāmsā, Tantra, Purāṇa, Mathematics, Alchemy (Rasasāstra) etc. it rose into eminence, and the spread of the text of the Vṛkṣāyurveda and its survival in this process of cultural contact with the rest of India is a near possibility. Dependence on a single Ms can not be taken as a sign of weakness as the Bhelasamhitā has survived till now through a single Ms.

The present Ms is written in the Sūtra style with prose and verses elaborating the text. In the Puṣpāngasūtrīyādhyayah the Ms has been written in the manner of Kulaka (five verses making one group) from verse 63 onwards. The compilatory fashion of collecting verses at the end of almost all the chapters has been noticed with the headings "tatra ślokāh" and "bhavanti cātra slokah" et seqq. This is reminiscent of the style found in the texts of the Nāṭyaśāstra of Bharata and in the Arthaśāstra, the Caraka Samhiā and the Suśrutasamhitā. The definitions of the chapters have been given with the suffix ja=iya, a style which also marks the texts of both Caraka and Suśruta. The language used is classical Sanskrit, and is not very different from the language used in the Upaniṣads and in the Māhabhāṣya,-except the technical aspect of a treatise on Botany. The highly developed philosophical texts are all written in the most simple language. The language of the Māhabhāṣya is marked by lucidity and clarity of expression and this feature also marks the present text. But even with its simple linguistic structure the Ms contains words and expressions which suggest its basic antique character. The Ms is marked by lacuna, misplacement of some portions of the the text and the absorption of portions of other well known texts like that of Caraka which come up in the normal course of dialogue. To wit, the opening verse of the Ms echoes the text of Manu 1.8 "sisṛkṣur vividhāh prajāh" in a faulty manner and also the verse 9 (Bījotpattikāṇḍa, Ch. 1) echoes the text of Manu 1.49 "antaḥsaṃjnāḥ bhavantyete sukhaduhkhasamanvitāḥ". Archaic expressions like "yadi cet" can be seen in Dwigaṇīyādhyāya Bījotpattikāṇḍa, verse 25; and "api cet" in Aṣṭāngasūtrīyādhyāya, Bījotpattikāṇḍa, verse 6 which finds parallel in the Gītā IX. 40.

The text as stated herewith uses a number of obscure, recondite and uncommon words, some of which are not to be seen in the Amarakosá or in any other lexicons. Some are to be found in the texts on medicines only. Thus the word 'puplika' (Bījotpattikāṇḍa, Puṣpāngasūtrīyādhyāya, verse 38) with an uncertain meaning and the word 'kīkhosa' (Bījotpattikāṇḍa, Dwigaṇīyādhyīya, verse 4) indicating the outer crust of a seed are absent in all the lexicons. 'Vṛjina' in the expression 'vṛjinapatra' (Bījotpattikāṇḍa, Vṛkṣāngasutriyādhyāya, verse 50) has been used in a very rare sense meaning 'crooked' which is found only in the Śāśvatakośa, a work of the 8th cent. A.D.[11] 'Jrmbhita' (Bījotpattikāṇḍa, Puṣpāngasūtrīyādhyāya, verse 34) used in a very special sense to indicate the curvature of lion's jaw comparable to the outer form of certain flowers, is a peculiar use not traceable anywhere else. The word 'kañcuka' (Bījotpattikāṇḍa, Dwigaṇīyādhyīya, verse 5) has been used in a special Botanical sense in slight modification from its known literary meaning. Moreover, in the nomenclature of leaves (Bījotpattikāṇḍa, Vṛkṣāngasutriyādhyāya, verse 38-39) the Vedic words 'Juhu' and 'Sruba' have been applied to indicate the external features of some leaves. The use of these Vedic words as simile is a technique found in the Mahābhārata and in the Śiśupālavadam of Māgha.[12] The word 'kārava' (Bījotpattikāṇḍa, Puṣpāngasutriyādhyāya, verse 12) which in all probability refers to the spoke of an umbrella can be traced only in the texts of medicine like the Caraka Samhitā and the

11. Sasvatakosa, ed by E.D. Kulkarni, Oriental Book Agency, Poona 1929.

12. Dronaparvan Ch. 36 and Sisupalavadham 15. 52.

Suśruta Samhitā[13] in the particular sense.

The Ms evinces other astounding peculiarities too. In some places (Bījotpattikāṇḍa, Puṣpāngasūtrīyādhyāya, verse 59; Phálāngasūtrīyādhyāya, verse 21; Aṣṭāngasūtrīyādhyāya, verse 29) the text begins with the indeclinable "khalu" which is not permissible in literary use; literature retains a solitary instance of its use to indicate prohibition. The mixture of prose and verse which has been held by the European scholars as a sign of the anteriority of any work, is found in abundance in the texts of the Nāṭyaśāstra of Bharata, in the Arthaśāstra of Kautilya, in the texts of Caraka and of Suśruta, and in the texts on Nyāya of the early days of the Christian era. All these assign an antique flavour to the present Ms. The other antique features exhibited by the Ms include (i) reference to other authorities e.g.Bījotpattikāṇḍa, Dwigaṇīyādhyaya, verse 13, "Kaścidāha.......(ii) the form of addressing the disciples as seen in the Smṛti texts of earlier ages e.g. Vānaspatyakāṇḍa, Samivargādhyāya, verse 4, "tadeva vargabhedena tvihānuvakṣyate śṛnu".

It is interesting to note that the text makes a unique revealation stating that the plants under references will be referred to both by their local names and their chaste names (Vānaspatyakāṇḍa, Samivargādhyāya, verse 9). This is possible only after systematic study and observation extending over a long period of time. It is to be noted in this context that the use of the word 'vyuhyamānatvāt' (Bījotpattikāṇḍa, Aṣṭāngasūtrīyādhyāya, verse 13) referring to the functional aspect suggests the command of the author of the text over the use of language befitting a technical text on Botany. Similarly the correlated use of the word 'puṣpavant' meaning 'the sun and the moon' after Amara (kālavarga) to indicate a phenomenon of Botanical significance (Vānaspatyakāṇḍa, Vanaspatinirvarṇanādhyāya) shows the high degree of precision attained by the author of the text in expressing scientific

13. Susruta Samhita, Sutrasthana, Ch. 46.

concepts. Examination of the Ms shows some corrupt expressions at some places. For example, there is a wrong expression 'divāyām' for simple 'diva' noticed in Bījotpattikāṇḍa, Puṣpāngasūtrīyādhyāya, verse 52. The arrangement of verses 8-9, Bījotpattisūtrīyādhyāya (Bījotpattikāṇḍa) is found defective as the word 'puṣpaṁ phalaṁ' in separation gives little sense. The correct word should be 'puṣpaphalasamanvitah'. The subject of the verb 'nayet' is absent. If, however, the two verses are reconstructed as-

"atha kālaprakarṣeṇa puṣpaphalasamanvitah vṛkṣaḥ

sañjno bhavedantaḥ sukhaduhkhasamanvitah

bījam tu kālaparyayād udbhitvā pṛthivīm nayet"

the text will yield a correct and consistent meaning. The discoverer of the present Ms noted in pagees 14, 71, 100, 109 etc. as certain portions indistinct in the original copy suggesting thereby that the original Ms was defective.

These facts taken together lead to the conclusion that the original of this Ms as discovered by the late J.N. Sircar Viṣagratna might be very old, antedating the present one probably by about a century or more. And kernel of the text may go back to the earlier ages of the Christain era.

Lallanji Gopal[14] has observed that the references to the names Kālañjara, Drāvida, Utkala, Prayāga etc. in this text (Bījotpattikāṇḍa, Vanavargādhyāya) indicate a late date for the Ms. It may, however, be mentioned that the name of the Kālañjara hill occurs in the Mahābhārata[15]. The Drāvida country also known as Drāmida has been referred to along with Utkala in the Rāmāyana.[16] Drāvida has also been referred to in the Manusamhitā.[17] Prayāga has been mentioned both in

14. Aspects of History of Agriculture in Ancient India by Lallanji Gopal, Bharati Prakashan (Varanasi) 1980, p. 37.
15. Vanaparvan Ch. 85, Vss. 56-58.
16. Rāmāyana 4.41
17. Manusamhitā 10.44

the Rāmayāna[18] and the Manusamhitā.[19] Lallanji Gopal has further observed that the descriptions of the forests found in the present text are in fair agreement with the descriptions of the forests in the Mānasollāsa.[20] A close comparison shows that the description of the forests covers 14 verses in the Vṛkṣāyurveda while in the Mānasollāsa it covers only 8 verses; and the description is palpably abridged in the latter work. Only three verses i.e. 174, 175, 176 of the Mānasollāsa agree ***verbatim*** with vss 16, 17, 18 (in parts) of the Vṛkṣāyurveda. The Verse describing the Angireyavana is corrupt in the Vṛkṣāyurveda with lacuna (vs 15). The word 'Vangāla' appears in the Mānasollāsa vs 173 which is clearly of very late origin being as late as the 14th cent A.D. The Vṛkṣāyurveda gives the distribution of the natural forests in ancient India which is fully in keeping with the spirit of the text as a work on Botany but in the Mānasollāsa the whole thrust is on the habitat of the elephants. Thus the basic approach is altogether different in both the texts. The two texts might have drawn upon a common source but there is no reason to assume that the Vṛkṣāyurveda borrowed from the Mānasollāsa.

Examination of the text makes it clear that the author of the text had profound and varied knowledge of the flora of ancient India. The intimacy displayed with the soil characteristics, the atmospheric conditions of the land and its characteristic vegetation is found unique. The text further reveals the author's deep acquaintance with plant taxonomy, the various means of propagation, concepts of pollination and sexuality. The Aṣṭāngasūtrīyādhyāya,contains the most scientific exposition on some of the physiological processes in plants like transportation of fluid, assimilation of food etc., known so far through ancient texts of India.

---

18. vide "The genuineness of the Uttarakāṇḍa of the Rāmāyana" S. JanakiFel. Vol. 1991 published by the Kuppuswami Sastri Re search Inst. Madras.
19. Manusamhitā 2. 21.
20. Mānasollāsa, 1, pp. 44, vss. 172-179.

Instances can be cited further to illustrate that some basic concepts of plant science were known to the ancient Indians. Thus the Ṛgveda* says that human habitation should not be allowed to tread on vegetation. And Manu (4.73) has prohibited stay or walk under the trees at night. Furthermore in a long discourse in Ch. 177 of the Śāntiparvan, Mahābhārata questions have been raised about the existence of life in plants which the sage Bhṛgu tries to explain the establish as follows : Like the human body, the body of the plant is also made of five elements. The plants react when they are struck. The atoms making the body of the plants are closely stuck but still there is some space in between the atoms, which enable them to assimilate the rays of the sun, rain and air causing the blossoming of the flowers, and bearing of fruits. The trees have the capacity to suck water through roots. Withering of trees, creepers, and of their foliage from excessive heat or from thunder show that they have both the power of feeling and hearing.

These details, put through commonplace analogy and popular belief, highlighting basic concepts of plant life must have arrived at from systematic observation and analysis. It will not be out of place to add that Guṇaratna, a commentator of the Ṣaḍdarśanasamuccaya also referred to the existence of consciousness among plants in 1200 A.D., keeping alive the tradition of the Manusamhitā and the Mahābhārata.

To end this review I am inclined to state here that the basic text of the Vṛkṣāyurveda of Parāśara, as presented herewith, can be placed in all probability in the 3rd-4th cent A.D. when the Manusamhitā took its final shape. The critical review by the editors provides ample evidences that the present text includes a great deal of knowledge on plant science that survive in Indian tradition through the ages.

---

* 10, 146, 1-6.

It is laudable that Dr. Roma Sarkar has taken up the mantle from her illustrious father, the late N.N. Sircar (senior editor) in bringing this work to its completion in the form of the present edition. It is hoped that the text edited herewith will receive attention of scholars at home and abroad as an outstanding contribution in the field of Botany in ancient India.

B.D. 25 Railpukur Road
Deshbandhunagar
Calcutta-59, India
July 7, 1994.

(Dileep Kumar Kanjilal)
M.A., Ph. D (Cal), B. Litt (Oxon)
Principal
Sanskrit College,
Calcutta

Preface

We have edited here a manuscript the Vṛkṣāyurveda of Parāśara, which is a full-fledged treatise on the plant science in Sanskrit. The original manuscript was discovered by the late Vaidyaśāstri J.N. Sircar Viṣagratna, who was an Āyurvedic Physician in the erstwhile princely state of Coochbehar, Bengal India. The manuscript was discovered sometimes before 1928. The late J.N. Sircar referred to the above discovery to Dorothea Chaplin of England in one of his correspondences to her. Chaplin mentioned about the extant manuscript in her book "Some Aspects of Hindu Medical Treatment", published by Luzak & Co, England in 1930. J.N. Sircar made a hand written copy of the original, and translated it into Bengali with a commentary, with an endeavour to publish the work, which however did not materialise. After his death the manuscript came into notice of his son, the late N.N. Sircar, the senior editor in the present work, and who was my father. No record was found as to the location of the original manuscript. The only impression one can have from the copied version is that the original manuscript was found in a somewhat mutilated state as the copier left spaces blank at some places with the comment that the reading was indistinct. Also the last portion of the manuscript was found missing.

In 1950, N.N. Sircar presented an article "An Introduction to the Vṛkṣāyurveda of Parāśara" in the Monthly meeting of the Royal Asiatic Society of Bengal (1950, 4th April, Year-Book R.A.S.B. for 1950, vol. XVII p. 176) with extensive quotations from the original Sanskrit text. This was later published in the Journal of the Royal Asiatic Society of Bengal (1950, JASB, vol. XVI, No. 1, pp. 123-139). Some aspects of the above introductory account were reviewed by G.P. Majumdar in his article "The History of Botany and Allied Sciences in Ancient India" (AIHS, UNESCO, Paris, 1951). The Indian National Science Academy, New Delhi, published a book "A concise History of Science in India" edited by D.M. Bose, S.N. Sen and B.V. Subbarayappa (1971), where the contents of the present manuscript were reviewed in detail drawn upon the above articles.

In 1989, a project "Vṛkṣāyurveda with English translation and notes", sponsored by the Indian National Science Academy, New Delhi, was initiated by N.N. Sircar (Principal Investigator) and myself. Unfortunately, N.N. Sircar, my revered father, passed away shortly after the completion of the above project. Thus it rests upon me the responsibility to bring the work to its final form. It is a matter of pride for me to be instrumental in bringing to culmination of an effort that was initiated about 70 years back by the late J.N. Sircar, my grand father.

As regards the original manuscript or any other copy of the same, we are still continuing our search. The basic copy of the manuscript that was made by J.N. Sircar, and other related documents are at present in my possession, which I intend to present to any institute of Oriental Study of India.

1995 **Roma Sarkar**
India

Table of Transliteration

Sans.	*Eng.*		*Sans.*	*Eng.*	
अ	A	a	ट्	Ṭ	ṭ
आ	Ā	ā	ठ्	ṬH	ṭh
इ	I	i	ड्	Ḍ	ḍ
ई	Ī	ī	ढ्	ḌH	ḍh
उ	U	u	ण्	Ṇ	ṇ
ऊ	Ū	ū	त्	T	t
ऋ	Ṛ	ṛ	थ्	TH	th
ए	E	e	द्	D	d
ओ	O	o	ध्	DH	dh
ऐ	AI	ai	न्	N	n
औ	AU	au	प्	P	p
°	Ṁ	ṁ	फ	PH	ph
°	Ḥ	ḥ	ब्	B	b
क	K	k	भ्	Bh	bh
ख	KH	kh	म्	M	m
ग	G	g	य्	Y	y
घ	GH	gh	र्	R	r
ङ	Ṅ	ṅ	ल्	L	l
च	C	c	व	V	v
छ	CH	ch	श	Ś	ś
ज	J	j	ष	Ṣ	ṣ
झ	JH	jh	स	S	s
नं	Ñ	ñ	ह	H	h

Chronology of the Sanskrit Works referred to

Ṛgveda	-	C. 1500 BC.
Atharvaveda	-	C. 1000 B.C.
Taittīrya Samhitā	-	C. 1000 B.C.
Vājasaneyi Samhitā	-	C. 1000 B.C.
Upaniṣad	-	C. 1000 B.C.
Mahābhārata	-	C. 400 B.C.
Caraka Samhitā	-	C. 4th cent B.C.
Arthaśāstra	-	C. 2nd cent. B.C. to Ist cent. A.D. (P.V.Kane)
Suśruta Samhitā	-	C. before 2nd cent. B.C.
Manu Samhitā	-	C. 200 B.C. to 200 A.D. (P.V.Kane)
Bṛhatsamhitā	-	C. Fifth cent. A.D.
Amarkośa	-	C. Sixth cent. A.D.
Dhanvantari Nighaṇṭu	-	C. Eighth cent. A.D.
Agnipurāṇa	-	C. Ninth cent. A.D. to Tenth cent. A.D.
Buddha Ghosa	-	C. Ninth cent. A.D.
Cakrapani	-	C. 1060-1100 A.D.
Guṇarātna	-	C. Eleventh cent. A.D.
Manasollasa	-	1129 A.D.
Vṛkṣāyurveda of Śūrapāla	-	C. Twelth cent. A.D.
Rājanighaṇṭu	-	C. 1235-50 A.D. (A.B. Keith)
Sārngdhārā Paddhati	-	C. Thirteenth cent. A.D.
Madanpāla Nighaṇṭu	-	C. Fourteenth cent. A.D.

Introduction

Vṛkṣāyurveda which literally means 'the science of plant life' is known to have existed in ancient India as a special branch of knowledge. Clear references to the Vṛkṣāyurveda is met with in the Arthaśāstra[1], the Brhatsamhitā[2] , and in the Agnipurāṇa[3], each having a section devoted to this branch of knowledge, generally dealing with the Agricultural practices; rules for planting trees - their proper seasons, and specific location for aesthetic and hygienic improvement of the homestead; as well as diseases of plants and their treatments. The Kāmasūtra[4] of Vātsyāyana includes the Vṛkṣāyurveda amongst the 64 arts developed in ancient India. There also exists a separate text on the Vṛkṣāyurveda attributed to Śurapāla, a work of later date (12th cent. A.D.) which bears an elaborate account only on the treatment of plant diseases. Further we come across a chapter entitled Upavana-Vinoda as a branch of the Vṛkṣāyurveda in the encyclopedic work compiled by Śārṅgdhāra (Śārṅgdhāra Paddhati, 13th cent. A.D.), which deals with construction and maintenance of gardens.

In contrast to these traditional texts or manuals of the Vṛkṣāyurveda mentioned above which mainly dwell on practical aspects of plant science, Parāśara's Vṛkṣāyurved, the

1. Ch. 24
2. Ch. 54
3. Ch. 282
4. 1. 3. 16

manuscript edited here is a full-fledged treatise covering some basic aspects of Botany. It contains detailed information on morphology and classification of plants, and other related topics. It is noteworthy that the plant diseases and their treatments which is one of the main topics of the traditional Vṛkṣāyurvedic texts as mentioned above, figures in the contents of the present text too. However this portion has been reported to be missing in the manuscript.[5]

This text is written in the classical 'Sūtra' (aphoristic) style which was a very old tradition, found largely in vogue in Sanskrit literature at the early centuries of the Christain era. The text is ascribed to Parāśara, a great sage, whose identity is difficult to establish. Earlier, Sircar (1950) held that he is the same Parāśara cited in the Caraka Samhita* as an authority on the science of Āyurveda. According to Sircar, Parasara wrote or compiled this treatise for the purpose of teaching Botany preparatory to the Āyurvedic studies. In favour of this view he argued that the work abounds in Botanical terminologies that could be found in the texts of Caraka and Suśruta as well as in many other Āyurvedic pharmacopoeas of later date. Majumdar (1951) accepted the above view and suggested that this work was compiled possibly between Ist cent. B.C. or Ist cent. A.D.

However, we have yet to come across any historical evidence that could corroborate the above view so far the identity of the author is concerned. It is yet to be established that 'Parāśara' referred to in the Caraka Samhitā is a historical personage rather than a mythical figure. As regards the date of compilation of the present text, the manuscript contents may provide us some useful clues. In this respect we would like to invite the attention of the reader to the review of Dr. D.K. Kanjilal in the Foreword of this book. He forms an opinion based upon the linguistic and other internal evidences of the manuscript that the kernel of this work goes back to a period as early as the 3rd-4th cent. A.D.

---

5. J.N. Sircar mentions in a note that the Cikitsita Kanda was not found in the original manuscript.

* C.S., Sutrasthana, Ch. i, 11

At this point, before we enter into details of the subject matter of the text it is relevant to have a glimpse of the traditional Indian knowledge of the plant science. Although we do not as yet know of any separate work dealing with the subject of Botany, still we get fragmentory information scattered in various traditional literature of India. These include the Vedas and the Upaniṣads, the Epics and the Purāṇas, the Buddhist and the Jaina Canonical works and the commentaries, the classical medical treatises like the Caraka Samhitā and the Suśruta Samhitā and their commentaries, the lexicon of Amarasimha as well as the many Āyurvedic Nighaṇṭus (lexicons or pharmacopoeas).

The Atharvaveda[1] is the earliest work which records some knowledge about morphology of plants. The Taittiriya[2] and the Vājasaneyī Samhitās[3] recognise different parts of a plant as Mūla (root), Tūla (shoot), Kāṇḍa (stem), Valśa (twig), Puṣpa (flower), Phala (fruit), Śākhā (branch) and Parṇa (leaf). The Caraka and the Suśruta Samhitā abound in various descriptive terms based upon the salient morphological traits of plants as their nomenclature. Nāmaliṅgānuśāsana of Amarsimha or the Amarkośa[4], a lexicographic work developed in the 6th cent. A.D. abounds in sound botanical information. Finally mention should be made of a class of Āyurvedic literature, called Nighaṇṭus or lexicons (pharmacopoeas) belonging to the period from 8th to 16th cent. A.D. (like Dhanvantari Nighaṇṭu, Madanapala Nighaṇṭu, Bhava Parāśara, Rāja Nighaṇṭu etc.) which also record valuable information about plants.

A critical review of the above materials of Botanical significance done by Majumdar (1927, 1951) clearly reveals how a systematic development of knowledge about plant and plant life took place in ancient India. Majumdar observed that the

---

1. AV., VII,7
2. Taitt. S. VII, 3. 19-20
3. Vāj. S. 22-28
4. Amarakośa, Vanauṣadhivarga

scientific study of plants in India formed the basis of allied sciences such as Medicine, Agriculture, Arbori-Horticulture etc. The importance of knowledge about plans in the science of Medicine is well attested in the Caraka Samhitā which states, "It is only a person who is well acquainted with the names and external features of plants, as well as able to use them properly according to their properties should be called an expert physician".*

Against this background we can now resort to a critical evaluation of the text material of the Vṛkṣāyurveda of Parāśara, and to find whether the knowledge embodied here has any traditional base. The text is divided into six parts (Kāṇḍa), viz., Bījotpattikāṇḍa, Vanaspatikanda, Vanaspatyakanda, Gulmakṣupakāṇḍa, Virudhavallikāṇḍa and Cikitsitakāṇḍa. The last part which as stated before is found to be missing in the original manuscript.

1. The first part, the Bījotpattikāṇḍa is again subdivided into eight chapters as follows :

Chapter I- The Bījotpattisūtrīyādhyāya contains a gen eral introduction to the subject matter and gives in brief an account of the origin of life.

Chapter II- The Bhūmivargasūtrīyādhyāya gives an ac count of the classification of lands accord ing to the soil and the climate with an outline of the flora characteristic to each type of land.

Chapter III- The Vanavargasūtrīyādḥyāya gives a distribu tion of forests in ancient India.

Chapter IV- The Vṛkṣāngasūtrīyādhyāya deals with the leaf and its various parts, and other leaf-like appendages; it also includes a description of the internal structure of the leaf.

* C.S., Sutrasthana Ch. 1, 56.

Chapter V- The Puṣpāngasūtrīyādhyāya describes the different parts of a flower; various types of flowers according to their shape and form; and also provides a system of classification ofplants based upon their floral characteritics.

Chapter VI- The Phalengasūtrīyādhyāya describes the fruit and its various parts; various kinds of fruits according to their shape, form and other morphological peculiarities.

Chapter VII- The Aṣṭāngasūtrīyādhyāya deals with the eight different plant parts viz., root, stem or trunk, heart-wood, sap, exudate, oleaginous substances and spines or prickles. The details of each of these plant parts as to their morphology or the physical properties or the functional roles have been furnished.

Chapter VIII-The Dwigaṇīyādhyāya gives an account of the seed and the seedling and the process of germination, and various methods of propagation of plants.

2. The Vānaspatikāṇḍa is divided into three chapters, and contains a general description of plants belonging to the group Vanaspati (cf. Moraceae family), and Tṛṇarāja (comprising the families Gramineae, Palmae, Araceae, Zingiberaceae and Liliaceae).

3. The Vāṇaspatikāṇḍa gives a descriptive account of the following families with special reference to their representative plant members.

(a) Cūtapuṣpa gaṇa (cf. Anacardiaceae)

(b) Devapuṣpa gaṇa (cf. Myrtaceae)

(c) Puplika gaṇa (cf. Rutaceae)

(d) Akṣapuṣpa gaṇa (cf. Combretaceae)

(e) Kuhapuṣpa gaṇa (cf. Rhamnaceae)

(f) Mallikāpuṣpa gaṇa (cf. Apocynaceae)

(g) Samivarga (cf. Leguminosae)

4. The Gulmaksupa kāṇḍa is a small chapter, dealing with the type of plants belonging to the Araceae family (Kaṇḍula gaṇa), which grow in moist and clayey soil; and also describes the morphological features of plants grouped as Simhapuṣpa gaṇa (cf. Acanthaceae), considered as example of shrub (Gulma)

5. The Vīrudhavallīkāṇḍa which seems to be incomplete, only enumerates twenty three different creepers.

The manuscipt here ends abruptly.

From the above account it is apparent that the information presented in the various chapters of this text, bear upon the following topics : Origin of life; Ecology; Distribution of forests; Morphology; Classification and Nomenclature; Histology and Physiology. What appears remarkable is that there exist in the text a great deal of information that are also found in the traditional literature of India, which we will set forth in the following discussion, considering each of the above topic separately.

Origin of Life

The text gives an exposition as to how the first living cell (Adi bīja) came into existence. It states that water transforms itself into a jelly like substance (Kalalaṁ) within which a dense organic mass (Piṇḍasthānukaṁ) is formed, which in course of time, being regulated (Vyuhan) by terrestrial energy (Vasunā) is converted into germ cell (Bīja). This appears to be an elaboration of the view that is found in the Upaniṣadic texts. The Garbhośārīroponiṣad states basically the same process in a simpler way that water attains a colloidal state (Kalalibhūta) and subsequently becomes a dense organic mass (Piṇḍa). In fact, the basic concept that life originates from water could be traced in some of the earlier Upaniṣads too, of which mention may be made of the Chāndogya Upaniṣad[1] which holds that the living or non living all are just water solidified.

Ecology

The classification of lands as given in this text is recognised to follow the same traditional approach as evident in the Caraka and Suśruta Samhitās.[2] It may be noted that the present text agrees well with the Caraka Samhitā; both delineate, in the same way, the physical environment typical for the Jāngala deśa and the Anūpa deśa. Further it agrees *ad verbatim* with the Caraka Samhitā in enlisting the flora of these two regions. The present text also incorporates various terms that are found in the Amarkośa[3] in relation to the character of soil like, Uṣavān (the soil is full of alkaline patches), Śarkarila (the land dominated by rock particles), Śikatila (the land dominated by sand), and Pankila (the land that remains muddy); and also various kinds of plant association like Swadwala (grass land), Nadwala (land dominated by reedy plants), Kumudwān (land abounding in lotus) and Vetaswān (land abounding in canes).

Distribution of Forests

The text provides us with some rare and valuable information as to the distribution of forests in ancient India. It gives the geographical boundaries of eleven natural forests that existed in India. It is worth-noting that the Mānasollāsa, an encyclopedic work, attributed to the Calukya king Somadeva (c.A.D. 1131) contains a description of eight of the above forests, as the places, where elephants abound in*. Curiously

1. Chānd. Up., VII, 10
2. C.S., Kalpasthāna Ch. 1,8-10; S.S., Sūtrasthāna, Ch. 35, 35
3. Amarakośa, Bhūmivarga

enough, the descriptions of these forests in both cases closely follow each other, even sharing some of the verses in common. For details, we refer to the Foreword of this book, which furnishes an elaborate and critical observation on this point.

Morphology

The present text contains a detailed morphological information on the various plant parts, evincing a remarkable accuracy of observation. It may be relevant to mention that the traditional literature of India also provide us with a great deal of information on plant morphology based upon minute observations. This kind of knowledge evidently developed from a need for precise identification of plants. Thus there came into existence a well developed system of nomenclature of plants based upon the salient morphological character, in ancient India. The names of plants that we come across in the various texts on Āyurveda starting from the Caraka and Suśruta Samhitās, bear ample testimony to the above system of nomenclature. Also it was a common technique in this system to use simile to depict particular features; similes used may be identified as follows : (i) any object like weapon (e.g. *Asipatra* - the leaf shaped like a sword, *Saccharum officinarum*, Rājanighaṇṭu); (ii) the overall shape of an animal (e.g. Maṇḍukaparṇī - the leaf resembling the shape of a frog, *Centella asiatica*, Caraka Samhitā); (iii) the various limbs of animals (e.g. Aśwakarṇa - the leaf shaped like the ear of a horse, *Shorea robusta*, Caraka Samhitā) and so forth. The same traditional approach is met with in the present text of the Vṛkṣāyurveda as evidenced by the vast terminology scattered throughout the body. It is noteworthy that many of these terms are found also in the traditional (vide Index).

It is interesting to note, that the set of plant organs that come under detailed consideration in this text are found

* Ch. 1, pp. 44, verses 172-179

exactly mentioned in the Caraka Samhitā[1], as plant parts for their medicinal importance. Further it appears significant that the descriptive account of the above plant parts available in the present text, bears information or ideas that are also known traditionally. Some important points are summarised below.

(a) Leaf - The external parts of a leaf are enumerated as Vṛntaṁ (petiole), Pakṣa (lamina), Patra Sirā (leaf-vein), Mārhi (rachis), Vistāra (tendril) and Paṭṭika (leaf-sheath). Among these, the use of the term Pakṣa, meaning the two lateral portions of a leaf (vide text) appears to be unique in this text, while the other terms are found in the traditional references conveying the same meaning, or in a slightly altered sense as in the case of the term Mārhi. This particular term in the present text is used to denote the rachis—the axis on which leaf-lets are borne in a compound leaf. The same term has been found in the Rājanighaṇṭu where it is used synonymous with Patra Sirā (leaf-vein)[2].

The text gives an exhaustive description of leaves according to their shapes, margins, surfaces and other characteristics. Most of the terms derived are found scattered in the Caraka and Suśruta Samhitās, the Amarakośa and the different Āyurvedic pharmacopoeas (vide Index). Further the technique of using simile, as stated before, is amply evident here. In addition difinitions of terms in this text agree well with their corresponding references elsewhere to identify plants. As for instance, Maṇḍukaparṇī is a term defined here as a leaf whose overall shape gives the impression of a frog. Traditionally, this term denotes the plant *Centella asiatica*, whose reni form leaves obviously fit the above description.

(b) Flower - The terms relating to the floral parts mentioned in this text, that are traditionally known too are Vallari (inflorescence), Vṛnta (pedicel), Dala (corolla), Keśara

1. C.S. Sūtrasthāna, Ch. 1, 42
2. The use of the term Mārhi in both cases may be justified and Botanically correct, as rachis (Mārhi) is part of the fibro-vascular system of the leaf which is actually the mid-rib (Patra Sirā).

(stamens), Parāga (pollen), Kinjalka (anther); while some terms which appear in this text with an extended sense are Jālaka for the calyx (the organ which covers the flower bud—vide text) and Varāṭaka, the style. In the traditional literature the term Jālaka refers to the flower bud (Rājanighaṇṭu) while Varāṭaka means a seed vessel (Amarakośa).

Of the different types of inflorescences mentioned in this text, the Chatrā (umbel) is a term which is met with in the various texts of Āyurveda to denote the group of plants belonging to the family Umbelliferae, where this particular type of inflorescence is found (Caraka Samhitā, Amarakośa, Dhanvantari Nighaṇṭu etc.). Also the term Kāravikā which comes under reference in this text, synonymous with Chatrā (Bījotpattikāṇḍa, Puṣpāngasūtrīyadhyāya, verse 12), is found in the traditional use to refer to the plants of the Umbelliferae (Kāravī, Suśruta Samhitā, Sūtrasthānā, Ch. 46, 240). Guccha Vallari is another term that is also available through the traditional sources (Rājanighaṇṭu). The various forms of flower (corolla) are described in this text in a typical archaic style of expression, as mentioned before; the bilabiate corolla is compared with the widely opened mouth (Jṛmbhitaṁ) of a lion. The flower of Vāsaka (*Adhatoda vasica*) which gives the above impression is traditionally named Simhāsya (Simha-lion, Āsya-mouth or jaw; Amarakośa). Similar such expressions found in the present text are Kapotvaktra-the corolla assumes the shape of the mouth of a pigeon; Nāsāpuṭa Samkāśam, the corolla resembling the nasal cavity; Pannagī Keśara, the anther resembling a horse-hoof and Urmikeśara where the anthers, united together, assume a wave like shape.

The present text records a fair knowledge of the internal organisation of an ovary. The ovary is recognised to be divided into chambers-Puṭaka, and to contain the placental tissue

* Actually the ovary in this case is unilocular but appears falsely trilocular due to projections of the placental tissue inwardly;

(Puṣa) inside. The two terms 'Puṣa' and 'Puṭa' can be related to the terms 'Trapuṣa' and 'Tripuṭa' respectively, which are known traditionally. The derivative meanings of the latter two terms seem to suggest that the internal structure of an ovary was noticed, even at an early period. The term Trapuṣa (Caraka Saṃhitā) denotes he plant *Cucumis sativa* (cucumber) and it means that the ovary or the fruit (ovary is later developed into fruit) consists of three placentas* (Tra-three, Puṣa-nourishing tissue or placenta). The term Tripuṭa (Dhanvantari Nighaṇṭu) which refers to the plant *Elettaria cardamomum* suggests that its ovary or fruit contains three locules or chambers (Tri-three, Puṭa-chambers or a hollow cavity) which is verified from actual observations.

Now at this point we would like to take into consideration some concepts presented in this text, concerning the function of the flower, which appears to be considerably advanced showing a proper scientic understanding. First of all we can refer here to the concept of sexuality. As to the traditional concept regarding sexuality of plants, we find in the Caraka Samhitā a vague and inaccurate notion with reference to the Vatsaka plant *(Holarrhena antidysenterica)*. It is considered that the plant which bears white flowers, large fruits and tender leaves should be male; while that which has red or yellow flowers, small fruit and small stalk should be female[1]. Such peculiarity of concept is found to be borne out in the Rājanighaṇṭu, a much later work, which gives a strange classification of plants into male, female and neuter based upon contrasting morphological attributes, such as slenderness or stoutness, softness or hardness, tallness or shortness etc. relating to the plants or their parts.[2] Curiously enough, in this text (Rājanighaṇṭu), we also find records of an actual observation as to the differentiation of sex in plants as in case of Ketaki[3] *(Pandanus odoratissimus)*. This plant is always referred to as a couple (Ketakidvaya). The male one is described as Dhūlipuṣpikā (flowers having dusts, i.e., pollen grains) and

Biphalā (unproductive of fruits). The female oen is known as Swarṇaketakī or Kanakaprasavā (that which yields golden harvest).

As regards the present text of the Vṛkṣāyurveda, we find a more clear concept of sexuality in plants, based on perfect scientific observations. It is recognised that a flower may be Saphalā (productive of fruits) or Niṣphalā (unproductive of fruits), attributed to the presence or absence of ovary respectively. The former is called Śrīpuspa or Raṇḍapuṣpa (female flower) and the latter is known as Mañjupuṣpa (male flower). It may be mentioned here that the phenomenon of unisexuality of flower in plants is found to be very conspicuous and easily noticeable in the Cucurbitaceae family (to which Trapuṣa, *Cucumis sativa* belongs). The present text also cites this family, which bears unisexual flowers (vide text, Tripuṣgaṇīya puspa, Bījotpattikāṇḍa, Ch. 5, verse 68). In this family flowers can be distinguished into two types clearly by their shapes; of which, one produces the fruit (female flower) while the other does not (male flower). Thus we can consider it to be quite natural that this kind of simple observations in nature might lay the basis for the development of the concept of sexuality as presented in the text of the Vṛkṣāyurveda of Parāśara.

As yet another advanced concept in this text, that merits attention, is the classification of plants based upon their floral characteristics. We will deal with this point later in the general context of classification.

(c) **Fruit** - Fruits are classified in various ways according to their nature, shape, name etc., and thus the text contains an exhaustive list mentioning different types of fruits, most of which are found available in the traditional texts on Āyurveda (vide Index).

(d) **Eight plant parts or Aṣṭānga** - As to the eight

1. C.S., Kalpathāna, Ch. 5, 4
2. Rājanighaṇṭu, Anupādi, Ch. 1
3. Rājanighaṇṭu, Karavirādi, Ch. 10

different plant parts (root, trunk or stem, bark, heart-wood, sap, exudates, oleaginous substances and spines or prickles), observations recorded in this text are often found to have a link with the traditional knowledge. Thus we can cite the following instances : Regarding bark, we come across the term Pāṭola, defined in this text as a type where the bark consists of several layers, placed one upon another. This term is found to be used traditionally (Dhanvantari Nighaṇṭu) to indicate the plant,—*Stereospermum suaveolens* which bears the above typical characteristic. Similarly as regards the stem, the present text describes a type- Pārvic, meaning that which has many joints. This type of observation is found recorded in the traditional texts as we come across the term Śataparva (stem having hundred joints) which is used as a synonym of *Bambusa sp,* which has this particular feature. Further the text refers to the term Sauṣira, where the stem is hollow inside; the term is used as another synonym of *Bambusa sp,* which typically bears this characteristic (Rājanighaṇṭu).

As for exudates (Niryāsa), the text makes a distinction according to their physical properties. Some of these types are known traditionally as we find various plants named after those characteristics. These are listed below. Kṣira - latex (Kṣīri or Kṣiravṛksa - Plant containing latex, referring to *Ficus bengalensis ;* (Dhanvantari Nighaṇṭu).

Moca - a kind of sticky exudate (Mocaniryāsaka, syn. Salmali, referring to *Salmalia malabarica;* (Dhanvantari Nighantu)

Jatuka - An inflammable substance of oleo-gum-resinous nature, which is also known as Kālaniryāsa or Guggulu (Dhanvantari Nighantu), referring to *Commiphora mukul.*

As to the Rasa or the sap that the plants produce, the text has an intersting discourse explained in the light of the Sāṃkhya doctrine, an old Indian philosophical system which holds that all gross objects are formed of the five primal elements—the Pañcamahābhūta, i.e. Kṣiti (earth), Ap (water), Teja (fire), Vāyu (air) and Ākāśa (space). This concept is also found to be integrated in the science of Āyurveda, the old Indian medical system. The present text, in dealing with this subject of Rasa, is found to represent the same basic idea what has been laid down in the Caraka Samhitā.* It states that a particular taste (Rasa) is the outcome of collocation of the above primal elements in various combinations and proportions. The various attributes or properties of each of the above primal elements are also enumerated.

(e) **Seeds and Seedlings** - The text also records detailed observations on seeds and seedlings as well as the process of multiplication. Various methods of propagation of plants other than by seeds have been outlined, Which may be either by stem cutting (Kāṇḍa) or by rhizomes (Kanda) or by root (Mūla) or also by leaves (Patra). It may be pointed out here, that these methods have also been known in India from the very early times. The process of stem cutting (Skandhabījam) is found mentioned in the Arthaśāstra[1] and the Bṛhat Samhitā[2]. The method of propagation by rhizomes or bulbs (Kandabījam) has been referred to in the Arthaśāstra. Further we get in Sumangalavilāsini[3] (a commentary on the Dīgha Nikāya) by Buddha Ghosha (9th cent. A.D.) description of various methods as Mula-bijam (by root seeds), Khandabīja (cutting), Phalu-bija (by joints), Agra-bija (budding) etc.

Also we come across in the present text of the Vṛkṣāyurveda some terms like Vṛkṣāruha and Abakeśi, which are also known traditionally (Amarakośa)[4]. Of the two terms (vide text) the former means the epiphytes, and the latter refers to the plants

---

* C.S., Sutrasthana, Ch. 26

1. Arthaśāstra, Ch. 24, English translation by Shama Sastri.
2. Bṛhatsamhita, Ch. 54, Vol II.
3. The Sumangala Vilāsinī - Buddha Ghosa's commentery on the Dīghanikāya, edited by Rhys Davids and Carpenter, P.T.S., Part I
4. Amarakośa, Vanausadhi Varga

which do not form seeds but reproduce through patrānkuras (cf. spores) developed on the dorsal surface of the leaves (cf. ferns). As for Abakeśi, the Amarakośa records the same basic idea, describing them as Aphala—unproductive or fruits.

Classification of Plants

In this text of the Vṛkṣāyurveda of Parāśara, primarily, we get a broad system of classification based upon the habit of plants and some other morphological peculiarities. Thus plants are classified into four broad groups as follows : Vānaspati, Vanaspatya, Virudha-Vallī and Gulma-Kṣupa (for explanation, vide text, Bījotpattikāṇḍa, Ch. 1, verse 11, 12). An almost similar classification is available in the Caraka[1] and Suśruta[2] Samhitās where the four groups are recognised as Vanaspati-trees which bear fruits but not flowers; Vanaspatya (Vrksa in the Suśruta Samhiā)- trees which bear both flowers and fruits; Virudhas-creepers, and Oṣādhi-herbs that wither away after ripening of fruits (cf. Annuals). According to Cakrapāṇi[3], the commentator of Caraka, the group Virudha comprises two classes, Latā (creepers) and Gulma (herbs), which agrees well with the treatment given in the present text.

According to the traditional system, the Vanaspati group refers to the various species of *Ficus*. Dalvana[1], the commentator of Susruta, cites examples of Vanaspati as Plakṣa *(Ficus talbot;)* and Udumvara *(Ficus recemosa)*. In actual observations, *Ficus* species have flowers borne on the internal wall of a closed receptacle (hypanthodium inflorescence). This basic idea, however, is found missing in the traditional description of Vanaspati (Caraka Samhitā, Suśruta Samhitā, Manu Samhitā) as stated above. It is interesting to note that the present text comes out with the clear observation on this point. It states that

---

1. C.S. Sūtrasthana, Ch. 1, 42
2. S.S. Sūtrasthāna, Ch. 1, 23
3. Positive sciences by the ancient Hindus by B.N. seal p. 169, Motilal Banarsidass, Reprint 1985

these plants bear flowers, hidden from sight (Gudha puṣpa), borne within a round and hollow thalamus (Phalgu-Sthālaka). The term Phalgu is used in this text as a common synonym for the *Ficus* species. It may be observed that this term is found to be an old usage, which refers similarly to the *Ficus* species as mentioned in the Caraka Samhitā (syn. Udumvara-*Ficus racemosa)* and in Dhanvantari Nighaṇṭu (syn. Kākodumvara-*Ficus hispida*). This seems to suggest that the knowledge about the internal organisation of the hypanthodium of *Ficus* might be in existence also at the time of Caraka, possibly at a technical level. Further the traditional usage like Krmiphala (Madanapala Nighaṇṭu) synonymous with Udumvara *(Ficus racemosa)*, which obviously refers to the presence of the insect (gall wasp—*Blastophaga*) inside the hypanthodium, points out that the internal structure of this organ came into notice of the ancient Indians.

Besides the broad classification as already mentioned, the present text provides an elaborate system of classification of plants based upon the morphological characteristics of flowers and fruits,—their similarities and dissimilarities. This is wholly a scientific approach to classify plants, based upon their natural relationships, conforming to the modern day concept of the systematic Botany. Again such a system of classification stands unique in comparison to the classification, laid down in the Caraka Samhitā and the Suśruta Samhitā, based upon a utilitarian approach (medicinal or dietic properties). It is however, difficult to prove on the basis of the available literary evidences, whether there was prevalent in ancient India any such scheme of classification, as presented in this text. But we find the basic concept of grouping of plants based upon the morphological similarities, developed already at an early period as is evident from the various stray references. For instance, the Amarakośa refers to the plants of Palmae (including Palm, Coconut, *Areca,* Date etc.) grouped under grasses (Tṛṇadruma), which is justified on the ground that similar to grasses, these plants are also characterised by seeds with one cotyledon, by presnce of spikes, and parallel veins on leaves. Further we notice that five different species of *Ficus* placed under a common group-Pañca Kṣira Vṛkṣa (a group of five plants exuding latex) which are Vaṭa *(Ficus bengalensis)*, Udumvara *(F. racemosa)*, Aśwattha *(F. religiosa)*, Parisa *(F. arnottiana)* and Plakṣa *(F. talboti)* as found in the Madanapāla Nighaṇṭu. Similarly the genus *Sida* may be mentioned which is traditionally identified as Balā, and under which four species are recognised (Balā Catustam) as Balā *(Sida cordifolia)*, Atibala (*S. rhombifolia)*, Mahābalā (*S. rhomboidea)* and Nāgabalā *(S. spinosa)*. As regards the system of nomenclature available in the present text, it is found quite in keeping with the traditional approach based upon the salient morphological feature of roots, leaves, flowers or fruits which have already been discussed in some details. The various descriptive synonyms for each individual plant member mentioned in this text (Vānaspatikāṇḍa, Vanaspatyakāṇḍa) appear also in the traditional literature.

Histology of the Leaf

The present text records a fair knowledge of the anatomy of the leaf. It is stated that there are innumerable cells in a leaf. They serve as a store house of the sap (Rasasyasca Ạdhārasca) that has got all the elementary properties (Pañcabhautic Guṇa) derived from the earth. Cells are microscopic (Aṇavasca) and contain a coloured substance (Rañjaka), and have cell wall (Kalābestitena). The cell wall is fine (Sūkhmapatrakā) which originates out of a jelly like substance (Kalalaṁ) under the influence of heat (Bhūtosmā). Such a description reminds us Caraka's observations of Kṛmies (worms) present in blood which states :.....Kṛmies (worms) are found in blood vessels (Sthānam Raktavāhinyo Dhamanyah), which are microscopic (Aṇu), circular (Vṛtta), without feet (Apāda), invisible on account of extreme fineness

(Sūkhmatwāccaika Bhavantyadṛśyah), of coppery colour (Varṇastāmrah).* Observations in both of the above cases, create the impression that in those days there must have been some instrumental devices that made it possible to carry out such anatomical studies.

Physiology

The text records a rudimentary concept relating to the transporting system (Sarvasrotāmsi) inside the plant. Of these, one that carries the Rasa (watery substances) from the soil to the different plant parts is called Syandanī, and the one that circulates the fluid both in the inward and outward directions is called Sirā. Thus the plant receives nourishment, and growth becomes possible. The text gives an elaborate description of the process (Bījotpattikāṇḍa, Ch. 7 vs. 13; Ch. 8, vs. 18) involved that the watery fluid drawn from the soil (Pārthivarasah) is transported from the roots to the leaves by the Syandanī. There it is digested in the presence of colouring matter (Rañjakena Pacyamānāt). It is interesting to note that some basic idea relating to the above process of assimilation of food can be traced in the Mahābhārata[1], where the following description has been given (rendered by Majumdar[2]): "Agni (energy) and air help in the digestion (assimilation) of the water (watery food materials) which is absorbed through roots of the trees. And it is on account of the assimilation of this watery solution that the vegetable kingdom undergoes development and becomes graceful".

From the foregoing account, it is quite clear that the text of the Vṛkṣāyurveda of Parāśara is predominantly based upon the traditional ideas prevalent in India on the subject of plant science. The vast terminology of Botanical significance that are available in the Āyurvedic literature are found embedded in this work. Also the text adheres to the many ideas that were

---

* S. Vimānasthānam, Ch. VII, 10

1. Mahābhārata, Śānti Parva, Ch. 184
2. Vanaspati by G.P. Majumdar, Calcutta University, 1927, p. 33

prevalent in Indian thought from very early period as that of the theory of the origin of life, the doctrine of Pañcamahābhūta (five primal elements), the classification of soils, the broad cllassification of plants, and a detailed system of nomenclature of plants based upon their Botanical charateristics. Again, what merits attention in this text, is the detailed and systematic treatment of the morphological characters of the plant members, and some advanced ideas specially concerning flowers. There may be doubt that the text may have some later accretions, because of the materials of advanced nature. However, the manuscript evidences do not favour such a view. Examination of the language in the Puṣpāngasūtrīyādhyāya, Bījotpattikāṇḍa, particularly the phraseology, vocabulary and the syntax reveals a continuity from the earlier chapters, suggesting that there is little possibility of insertion of any materials by a modern hand. On the other hand we may suggest that close observations of the natural phenomena and their logical analysis, independent of any instrumental aid, may also result in the development of advanced ideas as found in this text. It is also evident in the text that a great degree of accuracy comes in when the external features of plants are considered; on the contrary, knowledge appears to be largely deficient in ideas or theories when they are based upon mere speculation. As for instance, the text wrongly states that the primordial germ cell (Ādi bīja) develops, in the next stage, into the seed that gives rise to a complete plant (Bījotpattikāṇḍa, Ch. 1, vs. 6-7). This kind of lacuna is further reflected when the author of the text argues (Bījotpattikāṇḍa, Ch. 8, vs 13) as to which method of reproduction in plants is more advanced - the vegetative or the sexual. This definitely indicates a lack of knowledge about the complex evolutionery process from the unicellular organism to the more complex form of plants or animals. Similarly we can cite some other observations that appear incorrect in this text : Kanda or the bulb or the rhizome is identified here as a root which according to modern view is a modified stem. The text wrongly infers that the process of fertilization (Bījādānam) takes place inside the style (Bījotpattikāṇḍa, Puṣpāngasūtrīyādhyāya, vs. 39); but the process actually occurs inside the ovary.

We would like to end our observation with the hope that this treatise on plant science in Sanskrit, the Vṛkṣāyurveda of Parāśara will be subject to further review and discussion. At this point all we can say is - whoever might be the author or whenever the treatise might have been written, the work is certainly a unique production in the Sanskrit literature.

The technique of editing the text

In editing the present manuscript of the Vṛkṣāyurveda of Parāśara literary translation of the original Sanskrit text into English has been attempted with notes and examples wherever found necessary, given within parenthesis. In the English version topics have been provided with appropriate headings for the sake of convenience. The Sanskrit text has been presented here in its original corrupt form as far as possible. We would like to add here that due to technical difficulties it was not possible to use 'visarga' (h) and 'Anusvar' (m) uniformly in case of English transliteration.

In the Section of Illustration we have presented some photographs of the ink impression of original plant material, a technique devised by the late J.N. Sircar, for the purpose of plant documentation (Fig. 9 a-f).

Acknowledgement

The editors take this opportunity to express their deep and sincere gratitude to the memory of the late Professor G.P. Majumdar, FNI, who inspired and encouraged to a great extent the senior editor to edit this manuscript. The editors record sincere thanks to the Indian National Science Academy, New Delhi for sponsoring the project of editing and translating the manuscript of Parāśara's Vṛkṣāyurveda. Sincere thanks are due to Dr. D.K. Kanjilal, Principal, Sanskrit College, Calcutta for his help and valuable advice; to Dr.

Sunando Bose, formerly Professor in Botany, Bose Institute, Calcutta for his suggestions and comments; to Dr. A.K. Bag, Head History of Science Division, Indian National Science Academy, New Delhi for his interest and continued help and encouragement; to Ms. Sevantee Roy, Ms. Mohua Bose and to Mr. S.K. Chakravarty for the drawings; to Dr. U.C. Banerjee, Associate Professor in Biology, North Carolina Central University, Durham, NC, USA and to Dr. M.S. Bisht, Research Associate, G.B. Pant Institute of Himalayan Environment and Development for their helpful interest throughout this work and finally to Jingo for his inspiring presence.

Editors...

वृक्षायुर्वेदः

बीजोत्पत्तिकाण्डः ।

प्रथमोऽध्यायः

बीजोत्पत्तिसूत्रीयः

अथातो बीजोत्पत्तिसूत्रीयमध्यायं व्याख्यास्याम इति ह स्माह पराशरः ।।१ ।।

ब्रह्मास्वयम्भुर्भगवान सिसृक्षु विविध प्रजाः * ।
अग्रे ससर्ज स्थावरात्मानो वृक्षवीरुधः ।।२ ।।
एते क्षितिरुहाः सर्वे जगतो हितकारकाः ।
वने चैत्ररथे रम्ये समेताः पुण्यकर्मणः ।
भरद्वाजादयः सर्वे ये चान्ये मुनिसत्तमाः ।
ज्ञातुमिच्छन्ताः पप्रच्छूः पराशरं महर्षयः ।।३ ।।
पृष्टेण मुनिभिः सर्वैः पराशरोऽ व्रवीत्ततः ।
अथर्वाङ्गं प्रवक्ष्यामि ब्रह्मोक्तं वृक्षवैद्यकम् ।।४ ।।
उत्पत्तिं प्रकृतिभेदमाकृतिं भूमिभेदतः ।
वृक्षाणाञ्च चिकित्सितं यथाह चतुराननः ।।
वृक्षावल्लयस्तथा गुल्मा विविधतृणजातयः ।
ऋतुपुष्पफलान्ता वै ओषध्यो जज्ञिरे त्विह ।।५ ।।
तत्रादौ बीजमुत्पत्तिं श्रृणु मे दुर्लभं वचः ।
आपो हि कललं भूत्वा यत् पिण्डस्थाणुकं भवेत् ।
तदेवं व्युह्यमानत्वात् बीजत्वमधिगच्छति ।।
तत्र सिक्ता जलैर्भूमिरन्तरुष्माविपाचितम् ।
वसुना व्युह्यामानात्तु बीजत्वं प्रतिपादयते ।।६ ।।

---

* Manusamhitā, 1.7

तथा व्यक्तानि बीजानि संसिक्तान्यम्भसा पुनः ।
उच्छूजत्वं मृदुत्वं च मूलभावं प्रयाति च ।।७।।
तन्मूलादंकुरोत्पत्तिरंकुरात् पर्णसम्भवः ।
पर्णात्मकं ततः काण्डः काण्डाच्च प्रसरं पुनः ।।
अथ कालप्रकर्षेण पुष्पं फलं समन्वितम् ।
बीजम्तु कालपर्यायाद् उद्भित्त्वा पृथिवीं नयेत् ।।८।।
वृक्षः सञ्ज्ञो भवेदन्तः सुखदुःखसमन्वितः * ।
उद्भिज्जमाहुरुदभेदात् वृक्षस्तु द्विजसत्तमाः ।।९।।
द्रुमः क्षितिरुहः शाखी विटपी पादपस्तरुः ।
अनोकहः कुठः शालः पर्यायैर्वृक्ष उच्यते ।।१०।।
वनस्पतिर्वानस्पत्यो वल्ली गुल्मश्चतुर्विधः ।
तेषान्तु फलपाकान्ता त्वोषधिः प्रोच्यते बुधैः ।।११।।
अपुष्पवन्तपुष्पेण ते वनस्पतयः स्मृताः ।
पुष्पिनः फलिनश्चैव वानस्पत्या उदाहृताः ।।
वीरुद्वल्ली प्रताना स्यात् गुल्मास्तु तद्विवर्जितः ।
ह्रस्वशाखशिफः क्षुपो गुल्मस्तु परिकीर्तितः ।।१२।।
एतेषां क्षितिरुहाणां केचिद् बीजे प्ररोहति ।
काण्डे च सम्भवेत् केचित् तथैव पत्रकन्दयोः ।।१३।।
पत्रं पुष्पं फलं मूलं त्वक् काण्डः सारस्वरसौ ।
निर्यासस्नेहकण्टकबीजप्ररोहा एव च ।
इत्युक्तैर्गणसंग्रहैः प्रकृतिश्चादिशेत्तरोः ।।१४।।
वनस्पतिस्तथा काण्डः वानस्पत्यस्तथापरौ ।
वीरुद्वल्ली गुल्मक्षुपौ स्वे स्वे काण्डे प्रवक्ष्यते ।।
रोगलिंगं समुद्दिष्टं वृक्षाणां सात्म्यभेषजम् ।
सहव्यापदकारणं सर्वमेतच्चिकित्सिते ।।१५।।
अत्रैव परिकीर्तितं वृक्षायुर्वेदसम्मतम् ।
इति सूत्रं बीजोत्पत्तेर्ब्रह्माणा परिकल्पितम् ।।१६।।

इति पराशरकृते वृक्षायुर्वेदे बीजोत्पत्तिकाण्डे बीजोत्पत्तिसूत्रीयोनाम प्रथमोऽध्यायः ।

* Manusamhitā, 1.49

VṚKṢĀYURVEDA
Bījotpatti Kāṇḍa
Chapter - I
Bījotpattisūtrīyādhyāya

I will now explain the 'Bījotpattisūtrīya' - the chapter on the genesis of the organic life — thus said Parāśara [Verse 1]. Brahmā, the God omniscient in his desire to bring diverse kinds of animates created first the immobile forms - trees, shrubs. [Verse 2].

In order to acquire knoweldge about the plants beneficial to the world, performers of holy rites like Bharadwāja and other foremost of the sages assembled in the beautiful 'Caitraratha' forest (in the Himalayas) and questioned Parāśara (to enlighten them). [Verse 3].

When requested thus by the sages, Parāśara spoke (to the congregation): I shall narrate to you the Vṛkṣavaidyaka (Vṛkṣāyurveda) embodied in the Atharvaveda, and as revealed by Brahmā, the creator. [Verse 4]. The genesis of the plants, their nature and form as influenced by variation of land characteristics along with their care and treatment, and delineation of the essential features of the trees, shrubs, creepers, grasses, seasonal flowers (Ṛtu puṣpa) and annual plants (Oṣadhi) on the surface of the earth - what all the lord Caturanon* told me I shall now narrate the same. [Verse 5]

* An epithet of Brahmā

Now hear me delivering these precious words as to how organic life orginated. When water transformed into a jelly like substance ('Kalalaṁ'), within it was eventually formed a dense organic mass called 'Piṇḍasthānuka'. Then by a regulated process, it attained the nature of a germ cell. Subsequently, it went through metabolic changes induced by contact with water and by heat released from the soil. Thus regulated by terrestrial energies it was converted into a primordial germ cell (Ādi bījaṁ) [Verse 6].

The seed when moistened swells and becomes soft, leading to the initiation of the root. [Verse 7]. The root develops into a seedling which, in turn, bears leaves which nourish the stem. Then the branches come out from the stem. Thus, in course of time, the plant bears flowers and fruits. Seeds, in due course, pierces upwards through the soil. [Verse 8].

Plants have consciousness and are capable of feeling the sense of pleasure and pain ('Sukhadukha samanwitah'). The best of the 'twice born' calls plants 'udbhijja' because of their piercing upwards. [Verse 9].

'Druma', 'Kṣitiruha', 'Śākhī', 'Vitapī', 'Pādapa', 'Taru', 'Anokaha', 'Kuṭha' and 'Śāla' are all synonymous — denoting a plant. [Verse 10].

Plants are grouped into four classes, viz., 'Vanaspati', 'Vānaspatya', 'Virudha-Vallī', and 'Gulma'. Of these, those which die immediately after ripening of the fruits are called 'Oṣadhi' by the scholars (cf. Annuals). [Verse 11]. Plants whose flowers remain hidden (Apuṣpavanta) are called 'Vanaspati'. Such of the plants which bear visible flowers and fruits come under the class of 'Vānaspatya'.

'Virudha-Vallis' are the creepers having tendrils (Pratāna). 'Gulmas' or 'Kṣupas' have short branches and roots without any tendril. [Verse 12]

A number of these trees, creepers and shrubs reproduce from seeds. Some may be propagated through branches or cuttings. Some may grow from rhizomes or bulbs, and some other from leaves. [Verse 13]

Plants are characterised and classified into different 'Gaṇas' (cf. families) on the basis of their morphological characteristics of the various organs such as 'Patra' (Leaf), 'Puṣpa' (Flower), 'Phala' (Fruit), 'Mūla' (Root), 'Kāṇḍa (Stem), 'Sāra' (Heart-wood), 'Swarasa' (Sap), 'Niryāsa' (Exudation), 'Kaṇṭaka' (Spine and Prickle), 'Bīja' (Seed) and 'Praroha' (Seedling). [Verse 14]

The above group of plants, 'Vanaspati', 'Vānaspatya', Virudha-Vallī' and 'Gulma Kṣupa' each will be discussed separately in the respective section. Also there will be a chapter on plant diseases (Cikitsita Kāṇḍa) with aetiology (Kāraṇam) symptoms and proper treatments ('Sātmya bheṣajaṁ') indicated. [Verse 15].

What has been stated here conforms to the knowledge of the Vṛkṣāyurveda. Thus ends here the Bījotpattisūtrīyas as designed by Brahmā, the creatoṛ [Verse 16]. Here ends the first chapter, the Bījotpattisūtrīya of the Bījotpattikāṇḍa of the Vṛkṣāyurveda by Parāśara.

बीजोत्पत्तिकाण्डः
भूमिवर्गसूत्रीयाध्यायः

अथातो भूमिवर्गसूत्रीयमध्यायं व्याख्यास्याम इत्याह पराशरः ।।१।।

भूर्भूमिर्मृत्स्ना मृत्तिका क्षितिः क्षौणी वसुन्धरा ।
धरा धरित्री धरणिर्वसुधा चेति कीर्तिता ।।२।।
अत ऊर्ध्वं प्रवक्ष्यामि भूमिवर्गं स्वलक्षणम् ।
त्रिविधा प्रोच्यते भूमिर्जाङ्गलानुपमिश्रकैः ।।३।।
स्वल्पद्रुमो जाङ्गलश्च स्वल्पोदकमरुप्रायः ।
ऊषवन्तं शर्करिलः सिकतिलस्तथैव च ।।४।।
वृक्षा ये जायन्ते तत्र तेषां नाम विवक्षति ।
कदरखदिराशनोऽश्वकर्णधवतिनिसाः ।।
सल्लकीसालबदराः सोमवल्कश्च किंशुकः ।
आमलकवटाश्वत्थ समीककुभशिंशपाः ।।५।।
यवगोधुमवज्जिरसमीधान्यादयश्च हि ।
एतास्तु जाङ्गलजाता ओषध्यः परिकीर्तिताः ।।६।।
एवमेषां तुल्यप्रकृतीनां ये च वृक्षवल्लीक्षुपास्तेऽपि जायन्ते ।।७।
भूमेर्विरसभावत्वादग्निपवनपृथिव्यात्मकत्वाच्च जाङ्गले
नैसर्गिकरुक्षत्वाच्च प्रायशः कषाय
कटुतिक्तरुक्षरसान्विता वृक्षा उद्‌भवन्ति ।।८।।

अथानुप: पुन: सर: सरित् समुद्रपर्यन्त प्राय: । स्वाद्वलनड्वलकुमुदवेतस्वान् दल्वपङ्किलभूमिभाग:। सिक्ताम्बुधूम्रसंकाश-प्रवातबहुलो हिन्तालतमाल-कदलीनारिकेलवेत्रवेणुवल्वजवानीरवेतसोपशोभिततटतरङ्गिणीभि:। अनेक द्रुमवल्लीप्रतानातरुणविटपपुष्पितवनभूमिभाग: ।।९।। तस्मादानुपेभूमे: स्वरसभावत्वात् सलिलपृथिव्यात्मकत्वाच्च मधुराम्लप्रायोऽरुक्ष प्रकृतिवृक्षवल्ल्योषधिगुल्मा उद्भवन्ति सलिलस्य नैसर्गिकसौम्यत्वाच्च ।।१०।।

अथ मिश्रदेशो जाङ्गलानुपयोर्द्वयोर्देशयोर्भूमिलक्षणैश्च वृक्षवल्ली-गुल्मादिभिमिश्रदेशो व्यवस्येत् साधारणो वा ।।११।। तत्र मृत्स्ना पाण्डुरक्तकृष्णा वा नातिशुष्का नातिस्वरसा नातिसिकताशर्करान्विता भवति । तस्मात् मृत्स्ना साधारणत्वेन उर्वरा च सर्ववृक्षवल्लीगुल्मशस्याढयां भवति ।।१२।।

भवन्ति चात्र ।

ऊषवान् क्षारमृत्तिका शर्करिलस्तथा शृणु ।
त्रिविधा शर्करा विद्यात् लोहग्रावैका सम्मता ।।१३।।
द्वितीया सुधाशर्करा तृतीया प्रस्तरा मता ।
एताभि: संवृतदेशा: शर्करिल उदाहृत: ।।१४।।
सिकतिलश्च सैकतो मरुभूरुच्यते बुधै: ।
शद्वलो हरितशष्पैर्नड्वलो नडसंकुलै: ।।१५।।
कुमुद्वान् कुमुदप्रायैर्वेतस्वान् बहुवेतसै: ।।१६।।
चिरक्लिन्नञ्च भूमीनां दल्वञ्च परिकीर्तितम् ।
पङ्किल: कर्दमयुक्तैर्भूमिलक्षणमुच्यते ।।१७।।
इति पराशरकृते वृक्षायुर्वेदे बीजोत्पत्तिकाण्डे
भूमिवर्गसूत्रीयनाम द्वितीयोऽध्याय: ।।१८।।

Bijotpatti Kāṇḍa
Chapter - II
Bhūmivargasūtriyādhyāya

I will now explain the Bhūmivargasūtrīyādhyāya, the chapter dealing with lands – said Parāśara. [Verse 1].

The earth is variously termed as 'Bhū', 'Bhūmi', 'Mṛtsnā', 'Mṛttikā', 'Kṣiti', 'Kṣauṇī', 'Basundharā', 'Dharā', 'Dharitrī', 'Dharaṇi', 'Basudhā [Verse 2].

Now I shall discuss different types of land delineating their characteristic features.

Land is divided into three classes namely – 'Jāngala', 'Anūpa', and Miśradeśa'. [Verse 3].

Jāngala deśa: The tract is almost like a desert with scanty vegetation and limited water resources. The soil is sodic (Uṣavanta) with abundance of gravel and sand. [verse 4]. The plants which grow in such 'Jāngala deśa' are 'Kadara' (a kind of *Acacia* sp. with black heart-wood), 'Khadira' (*Acacia catechu*) 'Aśana' (*Terminalia tomentosa*), 'Aśwakarṇa' (*Shorea* sp.), 'Dhava' (*Anogeissus latifolia*), Tinisa' (*Ougeinia dalbergioides*), 'Sallakī' (*Boswellia serrata*), 'Śāla' (*Shorea robusta*), 'Badara' (*Zizyphus jujuba*), 'Somavalka' (*Ficus dalhousiae*), 'Kimśuka' (*Butea monosperma*), 'Āmalaka' (*Phyllanthus emblica*), 'Vaṭa' (*Ficus bengalenesis*), 'Aśwatha'

(*Ficus religiosa*), 'Samī' (some Leguminous species), 'Kakubha' (*Terminalia arjuna*), 'Śimśapā' (*Dalbergia sissoo*) [Verse 5]. 'Yava' (*Hordeum vulgare*), 'Godhuma' (*Trticium aestivum*), 'Bajjira' (*Pennisetum americanum*) and various types of legumes as well as the annual plants grow in this region. [Verse 6]. Besides these, there may be many other trees, shrubs and creepers having similar characteristics to those of the plants mentioned above, which also flourish in the region. Because of the extremely dry condition of the soil and natural arid environment, the elements of fire, air and earth prevailing in the biosphere result in producing plants containing astringent (Kaṣāya), pungent (Kaṭu) and bitter (Tikta) sap. [Verse 7-8].

Anūpa *deśa* : The area abounds in rivers, streams and lakes, and quite often sea happens to be the boundary. It has green grassy land and also has clusters of 'Nala' (*reedy plants*), Kumuda (*Nelumbo* sp.) and 'Vetasa' (*Calamus rotung*). The soil is clayey. Strong storm like wind laden with heavy moisture keeps blowing. Rows of 'Hintāla' (*Phoenix paludosa*), 'Tamāla' (*Garcinia morella*), 'Kadali' (*Musa sp.*), 'Nārikela' (*Cocos nucifera*), 'Veṇu' (*Bambusa* sp.), 'Valvaja' (?), 'Vānīra' (*Calamus Roxburghii*) and 'Vetasa' (*Calamus rotung*) bordering the river side present a scenic view. The forest land appears beautiful with the assemblage of various types of trees, shrubs with blossoming young branches [Verse 9].

Due to the moist nature of the soil and the influence of elements of water and earth as well as the inherent soothing effect of the water—herbs, creepers and annuals flourishing in the area generally bear sap that tastes sweet or sour. [Verse 10].

---

*. As an example of 'Anūpa deśa' the estuarine land of 'Sundarvan' of Bengal may be cited. It may be noted that in ancient India 'Sundarvan' covered all the estuarine districts of undivided Bengal including land covered by the present city of Calcutta.

Miśra deśa : The area where mixed features of Jāngala and Anūpa deśa are visible along with the vegetation of the regions. [Verse 11]. The soil is grey, red or black in colour. It is neither too dry nor too moist; neither it has an abundance of rock particles or sand. The land is fertile sustaining all kinds of trees, shrubs and crops. [Verse 12].

Summary

A land is termed 'Uṣavān' when the soil is full of alkaline patches. [Verse 13]. So also a land is called 'Śarkarila' when it is full of rock particles.

Rocks may be of three types :

mixed with iron ores.
composed of lime stone.
containing mixture of gravels, pebbles etc.

The tract covered by these rock types is called 'Śarkarila' [Verse 14]. The land dominated mainly by sand is termed as 'Sikatila'. The scholars also give the term 'Marubhū'. The land covered by green grasses all over is known as 'Swadwala' (grass land), and that dominated by Nala (reedy plants), Kumuda (lotus), Vetra (canes) are respectively called 'Nadwala', 'Kumudwān' and 'Vetaswān'. Where the soil ever remains wet and heavy throughout the year it is identified as the 'Dalwa-bhūmi' and that which remains muddy is the 'Pankila'. [Verse 15-17].

Thus ends here the Bhūmivargasūtrīyādhyāya, the second chapter of the Vṛkṣāyurveda by Parāśara. [Verse 18].

बीजोत्पत्तिकाण्डः

वनवर्गसूत्रीयाध्यायः

अथातो वनवर्गसूत्रीयमध्यायं व्याख्यास्याम इत्याह पराशरः ।।१।।

अटवी विपिनं चैव गहनं काननं वनम् ।
महारण्यमरण्यानी पर्यायैर्वनमुच्यते ।।२।।
वनानि सम्प्रवक्ष्यन्ते श्रूयतां द्विजसत्तम ।
निसर्गाद यत्र जायन्ते वृक्षवल्लीतृणादयः ।।३।।
यस्मिन् जनपदाश्रिते यद् वनं परिकीर्तितम् ।
प्रख्याते भारतखण्डे त्विह तदनुवक्ष्यते ।।४।।
वनं चैत्ररथं रम्यं देवगन्धर्वसेवितम् ।
कालकं किरातश्चैव महारण्यं प्रकीर्तितम् ।।५।।
वनं पाञ्चनदं प्राच्यं वेदिकारुषकं ततः ।
आङ्गिरेयं कालिङ्गेयं दाशार्णकं महाटवी ।।६।।
अपरान्तं सौराष्ट्रञ्च वनमित्यभिधियते ।।७।।
वनं चैत्ररथं रम्यं मानससरः शोभितम् ।।८।।
ततश्च प्रतीचीदेशे कालकं वनमुच्यते ।।९।।
प्राग्देशे किरातश्चैव ह्लादिनीप्लाविते स्थितम् ।।१०।।
त्रयमेतद् महारण्यं हिमाद्रिशिखराश्रितम् ।।११।।
सिन्धुसागरसङ्गमात् हिमालयकृतावधि ।

कालञ्जरे कुरुक्षेत्रे वनं पाञ्चनदं स्मृतम् ।।१२।।
गङ्गासङ्गमप्रयागहिमाद्रीणाञ्च मध्यतः ।
वनं प्राच्यमिति ज्ञेयं मध्यमं परिकीर्तितम् ।।१३।।
त्रिपुरकोसलादौ च वेदिकारुषकं वनम् ।।१४।।
उत्कले (Lacuna exists) वङ्गे चाङ्गिरेयं वनं स्मृतम्
। । १ ५ । ।
विन्ध्याद्रिचित्रकुटाद्रि कलिङ्गद्राविडाश्रितम् ।
वनं कालिङ्गाकं ज्ञेयं समुद्रावधि कीर्त्यते ।।१६।।
श्रीशैले वेदशैले च मलयाद्रौ तथैव च।
गन्धसारो भवेद्यत्र दाशार्णकं तदुच्यते ।।१७।।
सह्याद्रिभृगुकच्छान्तमपरान्तं वनं स्मृतम ।।१८।।
अवन्त्यां द्वारवत्यां च सौराष्ट्रवनमुच्यते ।।१९।।
एवमुक्तानि वनानि प्रकृत्या स्थानभेदतः
एतैर्भूम्यनुभवैस्तु वृक्षाणां प्रकृतिर्भवेत् ।
यथर्तुकालवैगुण्यात्तदेव चान्यथा स्मृताः ।।२०।।

इति पराशरकृते वृक्षायुर्वेदे बीजोत्पत्तिकाण्डे वनवर्गसूत्रीयनाम तृतीयोऽध्यायः ।।२१।।

Bījotpatti Kāṇḍa
Chapter - III
Vanavargasūtrīyādhyāya

Parāśara said : I will now narrate the Vanavargasūtrīyāni — the chapter dealing with forest regions. [Verse 1]. Forests are connoted by different terms like 'Aṭavī', 'Bipina', 'Gahana', 'Kānana', 'Vana', 'Mahāraṇya', 'Araṇyānī.', etc. [Verse 2].

Oh thou best of the twice-born ones, hear me describe the forests where trees, shrubs, creepers and grasses grow naturally. [Verse 3] I will describe the forests located in the different kingdoms (Janapada) within the famous land Bhāratkhaṇḍa (Land of India). [Verse 4]

'Caitraratha' vana : The beautiful sylvan tract frequented by Devas and Gandharvas (Gods and heavenly persons).

The dense 'Kālaka', and

'Kirāta',

'Pāñcanada',

'Prācya',

'Vedikāruṣaka',

'Āngireya',

'Kālingeya',

'Dāśarṇaka'
'Aparānta' and
'Saurāṣtra' vana [Verse 5-7].

In these forests, Kālaka vana is situated on the west of the Caitraratha vāna, adorned with the Mānsarovara. [Verse 8-9]. To the eastern side lies Kirāta vana with the river Hlādinī (Brahmaputra) flowing through. [Verse 10]. These three forests are located in the high altitude of the Himalayas. [Verse 11]. The forest Pāñcanada is spread over the land extending from the mouth of the river Sindhu (Indus) upto the Himalayas, including the region of Kālañjara and Kurukṣetra [Verse 12]. Prācya vana extends from Prayāg, the confluence of the Gangā (with the Yamunā), upto the Himalayan region. This is a medium sized forest [Verse 13]. Vedikāruṣaka vana spreads in Tripura* and Kosala [Verse 14]. Āngireya vana covers the neighbouring region of Utkala and Banga [Verse 15]. Kālingaka vana in Vindhyān and Citrakuṭa hills, extends further southwards, through the land of Kalinga and Drāvida, up to the sea coast (bay of Bengal). [Verse 16]. The forest in the vicinity of the hills 'Śrīśaila, Vedaśaila and Malaya Parvata is known as Dāśārṇaka vana where sandal wood (Gandhasāra) trees grow. [Verse 17].

The Aparānta vana has its location in the Sahyādri hill region spreading up to Bhṛgu - Kaccha (the Kutch of Gujarat). [Verse 18]. Saurāṣtra vana belongs to Avanti and Dwaravati. [Verse 19]. These woodlands, characterized according to their location and natural surroundings, are described above. The vegetation in these forests is influenced by the soil characteristics; variation may occur with diversity of seasons. [Verse 20]. Thus ends the Vanavargasūtrīyādhyāya, the third chapter of the Bījotpatti Kāṇḍa of the Vṛksāyurveda by Parāśara. [Verse 21].

---

* Tripura - Modern Tewar in the Jabbalpur district.

Distribution Of Forest Regions In Ancient India As Mentioned In The Vṛkṣāyurveda Of Parāśara

(Approximate Areas)

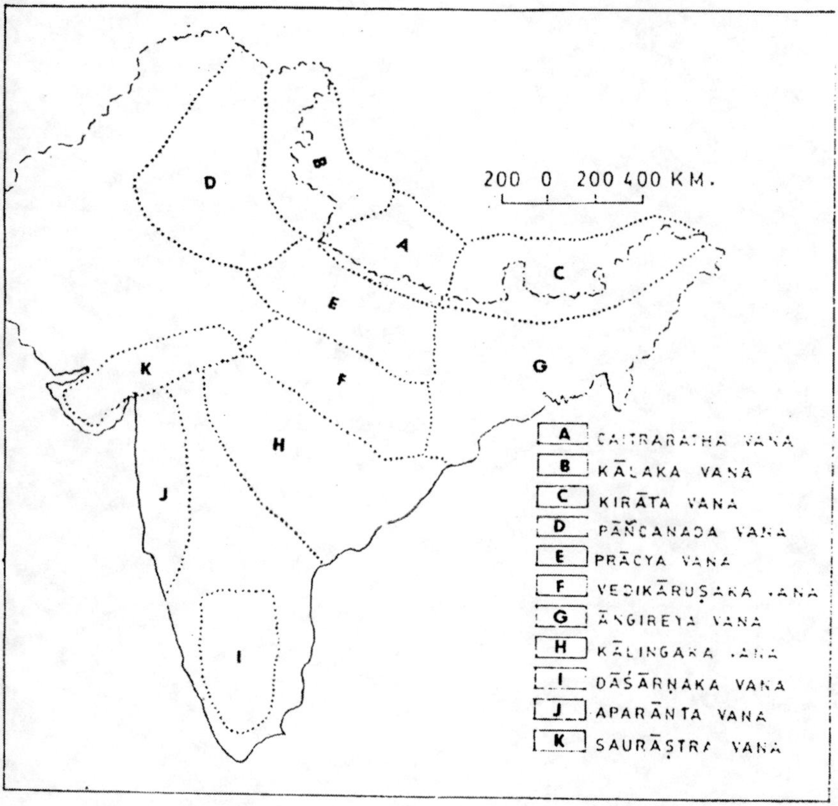

बीजोत्पत्तिकाण्डः
वृक्षाङ्गसूत्रीयाध्यायः ।

अथातो वृक्षाङ्गसूत्रीयमध्यायं व्याख्यास्याम इत्याह पराशरः ।।१।।

इह खलु सर्वेषां वृक्षवल्लीगुल्मानामङ्गप्रत्यङ्गैश्च गणसंग्रहो यथोद्देशमभिनिर्देक्ष्यामः ।।२।। तत्र वृक्षवल्लीगुल्मानां पत्रपुष्पफलमूलत्वक्काण्डसारस्वरसनिर्यासस्नेहकण्टकबीजप्ररोहश्चेत्यङ्गानि भवन्ति ।।३।। अङ्गैश्चेतैश्च वृक्षवल्लीगुल्मानां साधर्म्यं वैधर्म्यं तुल्यातुल्यप्रकृतिमभिसमीक्ष्य गणविभागमुपदेक्ष्यामः ।।४।। पुनश्चक्षित्यप्तेजोवाय्वाकाशत्मकानि यानि सूक्ष्माणि शरीराणि तान्युत्तरकालमुपदेक्ष्यते ।।५।। तत्रादौ पत्रमधिकृत्य गणाधिकारः । यथा पत्रं पर्णं पक्षं दलं पलाशं छदो बर्हमित्यनर्थान्तरम् ।।६।। नवपत्रन्तु पुनः प्रवालः किशलय इत्युच्यते ।।७।। पत्राणि तु वातातपरञ्जकान्यभिगृहणन्ति ।।८।। शिशिरे ये च पादपा जीर्णपत्रवर्जितास्तेषां माधवे पुनः प्रवालपल्लवानि समुद्भवन्ति । माधवे सञ्जातत्वाद्म-धुपर्णमित्यभिधियते ।।९।।

अथ खलु पत्रपक्षं वृन्तं पत्रसिरा रसकोषश्च माढिः च विस्तारञ्च पट्टिकञ्चेति पत्राङ्गानि भवन्ति ।।१०।।

अथ पत्रबन्धनन्तु सिरानां संसक्तेन वृन्ते यत् पत्रसंयोगम् । तच्च द्विविधं पृष्ठग्रन्थिकं प्रान्तग्रन्थिकञ्च ।।११।। पृष्ठग्रन्थिकबन्धनन्तु पत्रपृष्ठे यद् वृन्तसंयोगम् । प्रान्तग्रन्थिकबन्धनमेव पत्रपर्यन्ते यद् वृत्तसंयोगम् ।।१२।।

पत्रपक्षन्तु वृन्तादूर्ध्वमुभयतो यदङ्गम् ।।१३।। ततो वृन्तं पत्रस्य धारकं पुष्पफलयोरपि। कस्यचित् पत्रवृन्तमुभयतः पक्षसंकाशमङ्ग दृश्यते तदेव वृन्तपक्षमाचक्षते ।।१४।।

वृक्षवल्लीगुल्मानां शाखामाढिविस्तारेषु खलु सप्तविधं वृन्तबन्धनं विद्यात् । यथा संकुलपङ्क्ति पक्षपङ्क्ति - व्यत्यासपक्षपङ्क्ति-पुङ्खारा कुर्चबन्धनञ्चेति ।।१५।।

तत्र संकुलबन्धनन्तु वृक्षाणां शाखासु वैषम्येण यद् वृन्तसंयोगम् । एवञ्च माढिणा ।।१६।।

उभयपार्श्वगे शाखाविस्तारवल्लयादीनां वृन्तेन श्रेण्या यद् वृन्तबन्धनं तदेव पङ्क्तिबन्धनम् ।।१७।। पक्षपङ्क्तिकत्तूभयपार्श्वगे शाखाविस्तार-वल्लयादीनां समपक्षवृन्ताभ्यां श्रेण्या यद् बन्धनम् ।।१८।।

व्यत्यासपक्षपङ्क्तिकञ्च समपक्षवृन्ताभ्यां श्रेण्या व्यत्यासेन यदुपर्युपरि बन्धनम् ।।१९।। इषुपुङ्खाकारेण समपक्षपङ्क्तिकेन माढिनि यद् वृन्तबन्धनम् तदेव पुङ्खबन्धनम् ।।२०।।

अराबन्धनन्तु चक्रस्यारका इव शाखासु वृत्तमण्डलेन चक्राकारेण यदेव वृन्तसंयोगम् ।।२१।। कुर्चबन्धनञ्च कुर्च इव संस्थानं वृक्षशीर्षे यद्बन्धनम् । यथा तालनारिकेलक्रमुकखर्जुरेषु तृणजातिषु वृक्षेषु एव दृश्यते । इति सप्तविधं वृन्तबन्धनम् ।।२२।।

कस्मिंश्चिद् वृक्षे वृन्तहीनं पत्रं दृश्यते तच्च काण्डे संवीतं भवति। एतदेव काण्डाक्रान्तपत्रमाचक्षते ।।२३।।

पत्रसिरा तु पत्रे रेखाङ्कितेन दृष्टवती रसवहा च ।।२४।। पत्रे सिराणां सन्निवेशो द्विविधो भवति। प्रगुणं वेल्लितञ्च तत्र प्रगुणमृजुक्रमेण यत् संस्थानम् । वेल्लितन्तु वक्रेण वा सङ्कुलेन जालवद् यत् संस्थानम् । प्रगुणेन मौञ्जपत्रं वेल्लितेन तु जालिकपत्रं संज्ञायते ।।२५।।

प्रकृत्या पत्रपक्षं चतुर्विधं भवति । समपक्षञ्च विषमपक्षञ्च समकर्णिकञ्च विषमकर्णिकञ्चेति । तत्र समपक्षं खलु यदुभयपक्षं साम्यं भवति। विषमपक्षन्तु

यद् वैषम्येण विद्यते। समकर्णिकञ्च यत् पत्रपक्षं कर्णेन साम्यं भवति। विषमकर्णिकन्तु यत् पक्षं विषमकर्णेन विद्यते ।।२६।।

पत्रे रसकोषस्तु रसस्याश्रय आधारश्च ।।२७।। खलु वृक्षपत्रे रसकोषास्तूपरिसंख्येयाः सन्ति। ते कलावेष्टितेन पाञ्चभौतिकगुणसमन्वितस्य रसस्याशयश्च। एते रञ्जकयुक्तमणवश्च । कला तु सूक्ष्माच्छपत्रका या भूतोष्मापाचिता कललादुपजायते। आशय आधारश्चेति ।।२८।।

ततोमाढि यदेव गुच्छपर्णानां समिपर्णानाञ्च धारकं यच्च जीर्णपत्रेण सह प्रपतितं भवति । एतदेव द्विविधम् । एकसन्धिकं बहुसन्धिकञ्च । तत्रैकसन्धिकमेकेन सन्धिना सन्धितम् । बहुसन्धिकञ्च बहुभिः सन्धिभिश्चैव ।। कतिविधानां वल्लीजातिनां माढिसदृशमधिकमङ्गं दृश्यते यच्च शाखेव चिरेण काण्डे संसक्तं भवति तच्च विस्तारमाचक्षते । एव कस्यचित् शीर्षपत्रयुग्मौ तन्तुसंकाशौ भवतः । केषाञ्चिद् वल्लीनां पत्रवृन्तमूलात् सूत्राकारमङ्गं दृश्यते तच्च वितानमाचक्षते । आश्रयादौ तेन वितानेन तास्तु प्रताना भवन्ति ।।२९।।

सर्वेषु तृणवृक्षेषु पट्टिकं दृश्यते । एतेन पट्टिकेन सह तृणवल्वजानां पत्रं संसक्तं भवति । तच्च काण्डे काण्डाग्रे च संविशते । एव जीर्णपत्रेण सह प्रायशः प्रपतितं भवति । एतदेव खोल्लकं विज्ञायते ।।३०।।

केषाञ्चिद् वृक्षाणां नवपल्लवकोरकाच्छदकमधिकमङ्गं दृश्यते । तदेव शुङ्ग इत्याचक्षते ।।३१।।

केषाञ्चिद् वृक्षाणां पत्रवृन्तमुभयत उपपक्षं सञ्जायते ।।३२।।

खलु पत्रन्तु सामान्येन द्विविधं भवति। समपर्ण विषमपर्णं चेति । तत्र समपर्ण यदखण्डेन पत्रपक्षं साम्यं भवति। विषमपत्रन्तु यत् कर्णेन वैषम्यं भवति ।।३३।।

अत ऊर्ध्व वृक्षपत्रानां नामतो गणसंग्रहमुपदेक्ष्यते ।।३४।। तद् यथा- अर्घ्यपत्रञ्च जुहूपत्रञ्च स्रुवपत्रञ्च वृत्तबर्हञ्च मण्डलाग्रञ्च दीर्धदलञ्च वीरपत्रञ्च वाणबर्हञ्च शङ्कुबर्हञ्च सूचिपत्रञ्च इलिपर्णी च कर्णिकाछदश्च त्रिपर्णञ्च

क्रकचच्छदश्च वृजिनपत्रञ्च गुच्छपत्रञ्च श्रीपर्णी च सिम्बिपर्णी च युग्मपत्रञ्च नखपत्रञ्च बालपत्रञ्च करच्छदश्च मञ्जुबर्हञ्च चीरिपत्रञ्च कारवीदला च कचपत्रञ्च सूर्पपर्णञ्च विषमपक्षपर्णञ्च बलिपत्रञ्च चित्रपर्णी च पृश्निपर्णी च मण्डूकपर्णी च शुक्तिपर्णी च कमठच्छदश्चगोधापदी च गोकर्णी च अश्वकर्णी च हस्तिकर्णी च हस्तिपर्णिनी च हंसपदी च पारावतपदी च लोमशपर्णी च कण्टपर्णी च नाड़ीदला च पीवरीदला च मौञ्जपत्रञ्चेति ।।३५।।

ततः पत्रगणसूत्रीये नामतः समुद्दिष्टानि एते पत्रेगणाः स्वरूपतश्च यथोद्देशमभिनिर्दिश्यन्ते ।।३६।। येषां द्रुमवल्लीगुल्मानां पत्रमर्घ्यपात्राकारं पृष्ठग्रन्थिकञ्च तदर्घ्यपत्रगणीयम् ।।३७।।

येषां द्रुमवल्लीगुल्मानां पत्रं जुहूपात्राकारं प्रान्तग्रन्थिकञ्च दर्वीसंस्थानं वा तच्च जुहुपत्रगणीयम् ।।३८।।

यच्च स्रुवपात्रसंस्थानं प्रान्तग्रन्थिकं तदेव स्रुवपत्रगणीयम् ।।३९।।

वृत्तबर्हन्तु वृत्तमण्डलाकारं पृष्ठग्रन्थिकञ्चेति वृत्तबर्हगणीयम् ।।४०।।

मण्डलाग्रपत्रञ्च वृत्तमण्डलप्रायः पत्रशीर्षञ्च प्रान्तग्रन्थिकमिति मण्डलाग्रगणीयम् ।।४१।। दीर्घेण च प्रान्तग्रन्थिकेन दीर्घदलगणीयम् ।।४२।। वीरपत्रमसिवच्च प्रान्तग्रन्थिकञ्चेति वीरपत्रगणीयम् ।।४३।।

वाणबर्हञ्च त्रिकर्णिकं वाणफलकमिव प्रान्तग्रन्थिकञ्चेति वाणबर्हगणीयम् ।।४४।। शङ्कुबर्हञ्च शङ्कुनिभं प्रान्तग्रन्थिकञ्चेति शङ्कुबर्हगणीयम् ।।४५।। सूचिपर्णीगणीयं सूचिसदृशं भवति ।।४६।। इलिपर्णी तु करपालिकासदृशा चेति ।।४७।। बहुभिः कर्णयुक्तैश्च कर्णिकाछदगणीयम्। त्रिपर्णन्तु त्रिभिः पत्रैश्च प्रान्तग्रन्थिकमेवञ्च येषां पत्राणां पार्श्वसिराः पार्श्वाभितोऽभिसर्पितास्तदेव त्रिपर्णगणीयम् । केचिद् गुच्छपर्णमाह ।।४८।। करपत्रमिव क्रकचच्छदगणीयम् ।।४९।। ऊर्मिवत् कुञ्चितं वृजिनपत्रगणीयम् ।।५०।। पक्षैर्बहुभिर्युक्तैः प्रान्तग्रन्थिकमिति गुच्छपत्रगणीयम् ।।५१।। सूक्ष्माग्रं त्र्यश्रि तु वृन्तमभितो वृत्तमण्डलप्राय इति पत्रं श्रीपर्णगणीयम् ।।५२।। सिम्बिपर्णन्तु सुपर्णसङ्काशं त्रिभिः

पत्रैस्तेषामेकपर्ण मध्यतः श्रीपर्णसङ्काशं द्वौ पर्णौ माढ़िमुभयतः पक्षिपक्षसन्निभौ विषमपक्षौ तौ न्युब्जवृन्ताभ्यां सन्निबद्धौ । क्वचित् सिम्बिपर्णं बहुभिः समपक्षपत्रैस्तु माढ़िणि दृश्यते ।५३।। युग्मपत्रञ्च द्विखणिताग्रं वृन्तमभितोऽपि युगलपत्रमिव प्रान्तग्रन्थिकञ्चेति युग्मपत्रगणीयम् ।।५४।। बालपत्रञ्चातिक्षुद्रमायतञ्च शीर्षवृन्तमभितो वृत्तमण्डलप्राय इति बालपत्रगणीयम् ।५५।। नखपत्रञ्च नखसन्निभमिति नखपत्रगणीयम् ।।५६।। हस्ततलसङ्काशं करछदश्चेति करछदगणीयम् ।।५७।। मञ्जुवर्हञ्च नेत्राकृतिश्चेति मञ्जुबर्हगणीयम् । एतानि पत्राणि प्रान्तग्रन्थिकानि भवन्ति ।।५८।। चीरिपत्रञ्च झिल्लिकाया सदृशं विषमकर्णितं विभक्तं वं प्रान्तग्रन्थिकमिति चीरिपत्रगणीयम् ।।५९।। कारवीदला च केशराकारा विभक्तदलेति कारवीदलगणीयम् ।।६०। कचपत्रन्तु केशसदृशमिति कचपत्रगणीयम् ।।६१।। सूर्पपर्णन्तु वृहत्पत्रं सूर्पाकारञ्चेति सूर्पपर्णगणीयम् ।।६२।। विषमपक्षन्तु यस्य पर्णस्य पक्षं विषमं भवति तदेव विषमपक्षगणीयम् ।।६३।। बलिपर्णञ्च गजचर्मनिभञ्च वा खरपत्रञ्चेति बलिपर्णगणीयम् ।।६४।। चित्रपर्णञ्च विचित्ररेखाङ्कितञ्चेतिचित्रपर्ण गणीयम् ।।६५।। स्वलपायतं खरं तनुपत्रञ्च प्रान्तग्रन्थिकमिति पृश्नि पर्णगणीयम् ।।६६।। मण्डुकपर्णीतु मण्डुकसङ्काशमिति मण्डुकपर्णीगणीयम् ।।६७।। शुक्तिपर्णञ्च शुक्तिसंङ्काशमिति शुक्तिपर्णगणीयम् ।।६८।। कमठपर्णञ्च कमठाकारञ्चेति कमठपर्णगणीयम् ।।६९।। गोधापदीतु गोधापदाकारञ्चेति गोधापदीपत्रगणीयम् ।।७०। गोकर्णी तु गोकर्णसंङ्काशमिति गोकर्णपत्रगणीयम् ।।७१।। अश्वकर्णी तु अश्वकर्णसंङ्काशञ्चेचेति अश्वकर्णगणीयम् ।।७२।। हस्तिकर्णञ्च हस्तिकर्णसङ्काशमिति हस्तिकर्णगणीयम् ।।७३।। हस्तिपर्णिनी तु करिकरकुम्भसङ्काशमिति हस्तिपर्णगणीयम् पीलुपर्णी वा ।।७४।। हंसपदी तु हंसपदाकारञ्चेति हंसपदीपत्रगणीयम् ।।७५।। पारावतपदी च पारावतपदसदृशमिति पारावतपदीपत्रगणीयम् ।।७६।। लोमशपर्णञ्च लोमचितमिति लोमशपत्रगणीयम् ।।७७।। कण्टपत्रं च दुस्पर्शं कण्टकाचितमिति कण्टपत्रगणीयम् ।।७८।। नाड़ीपर्णञ्च नाड़ीसन्निभं सुषिरमिति नाड़ीपत्रगणीयम् ।।७९।। पीवरछदश्च

स्थुलपृथुलसान्द्ररसान्वितमिति पीवरपत्रगणीयम् ।।८०।। मौञ्जपत्रञ्च तृणवल्वजानां पत्रं येषां सिराणां सन्निवेशमधस्तादूर्ध्वमृजुव्यञ्जनमिति मौञ्जपत्रगणीयम् ।।८१।। पुष्पपत्रञ्च पुष्पाणामभिव्यञ्जकम् ।।८२।। असंख्येयानां वृक्षवल्लीगुल्मानां पत्रानामाकृतिनामाविस्कृततमा व्याख्याताः। अनेन विधिना चानुक्तेषु तदवयवं लक्षणञ्च व्यवस्येत् ।।८३।।

इति पराशरकृते वृक्षायुर्वेदे बीजोत्पत्तिकाण्डे वृक्षाङ्गसूत्रीय नाम चतुर्थोऽध्यायः ।।८४।।

Bījotpatti Kāṇḍa
Chapter - IV
Vṛkṣāngasūtrīyādhyāya

Parāśara said : I will now discourse upon the Vṛkṣāngasūtrīyas - dealing with the different organs of the plant [Verse 1]. I will deal with classification of trees, creepers and shrubs on the basis of their different organs. [Verse 2].

Parts of a Plant

Patra (Leaf), Puṣpa (Flower), Phala (Fruit), Mūla (Root), Twak (Bark), Kāṇḍa (Stem), Sāra (Hear-twood), Swarasa (Sap), Niryāsa (Exudation). Sneha (Oleaginous matter), Kaṇṭaka (Spine or Prickle), Bīja (Seed), Praroha (Seedling), are the different parts of a plant. [Verse 3].

I will give a classification of trees, shrubs, creepers by comparing the differences and similarities existing in their organs. [Verse 4].

Influence of Pāñcamahābhūta on plant life

Parāśara stated further that the effects of the elements of Pāñcamahābhūta viz. Kṣiti (Earthy elements),Ap (Aqueous elements), Teja (Solar energy), Vāyu (Air) and Ākāśa (Space) on plant life will be discussed later on. [Verse 5].

Leaf and its synonyms

Patraṁ, Parṇaṁ, Pakṣaṁ, Dalaṁ, Palāśaṁ, Chadaṁ and Barhaṁ are the various synonyms of a leaf. [Verse 6]. The young leaves are called Kiśalaya or Pravāla. [Verse 7].

The leaves take in air (Vāyu or Vāta), heat (Ātapa) and the colorific principles. [Verse 8].

Those plants which shed their leaves in winter, bear new leaves with the onset of spring. These foliages are termed 'Madhuparṇa' as they appear in Madhumāsa or Madhuṛtu (Spring). [Verse 9].

Different parts of a leaf

Patrapakṣaṁ (Leaf blade or lamina), Vṛntaṁ (Petiole), Patrasirā (Veins), Rasakoṣa (Cells), Mārhi (Rachis), Vistāra (Tendril), and Paṭṭika (Leaf sheath) are the different parts of a leaf. [Verse 10]. Fig. 1.

Patra bandhanaṁ (Articulation of the petiole)

The articulation of the petiole with the leaf lamina in conjunction with its veins is termed as Patra bandhanaṁ. It is of two different types, viz., Pṛṣṭhagranthika and Prāntagranthika. [Verse 11]. In the former case, the petiole is attached to the dorsal surface of the lamina, while in case of latter it is joined to the base of the lamina. [Verse 12].

Patrapakṣaṁ (Leaf lamina)

Patrapakṣaṁ refers to the two lateral portions of a leaf (cf. Leaf lamina or Leaf blade) above the petiole. [Verse 13].

Vṛntaṁ (Petiole/Pedicel/Peduncle)

Vṛtnaṁ is the part which holds the leaf, the flower and the fruit in space. In some leaves, Vṛntaṁ is associated with a laterally expanded wing on each side. These wings are termed Vṛnta Pakṣaṁ or Upapakṣaṁ (cf. Winged petiole). [Verse 14].

Vṛnta bandhanaṁ (Attachment of the petioles)

Vṛnta bandhanam refers to the arrangement of the petioles with the branches and the rachis. (cf. Phyllotaxy).

It may be of seven different types as observed in trees, creepers and shrubs, viz. Sankulapankti, Pakṣapankti, Vyatyāsapakṣapankti, Punkha bandhana, Arā bandhana, and Kurca bandhana. [Verse 15].

Sankula bandhana

The petiole and also the rachis are arranged on the branches in an irregular manner. [Verse 16].

Pankti bandhana

The petioles are arranged in rows on both sides of the stem. [Verse 17].

Pakṣapankti bandhana

Pairs of petioles are arranged successively in rows on both sides of the branch of a tree or a creeper (cf. Opposite superposed). [Verse 18].

Vyatyāsa pakṣapankti

Pairs of petioles are successively placed one upon the other, each pair being oppositely directed relative to the other (cf. Opposite decussate). [Verse 19].

Punkha bandhana

Pairs of petioles are arranged in rows on the rachis in a fashion resembling the feathered end of an arrow. [Verse 20].

Arā bandhana

The petioles are arranged in whorls around the stem like the spokes of a wheel (cf. Verticillate). [Verse 21].

Kurca bandhana

The leaves are borne on top of a tree trunk similar to that of a brush–head. This type of arrangement is commonly found in plants like palymra, coconut, date palm, betel palm and also in grasses. [Verse 22].

Kāṇḍākrānta Patra (Sessile leaf)

In some plants leaves do not have any petiole and their bases are attached to the stem forming a sheath around it. [Verse 23].

Patra sirā (Veins of a leaf)

The fine lines spread on the surface of the lamina are called Patrasirā (cf. Veins). They carry the sap. [Verse 24].

Sirā sanniveśa (Venation)

The venation of a leaf is of two types, viz. Praguṇa and Vellita. In Praguṇa type the veins run straight (Ṛjukrameṇa cf. Parallel venation). In Vellita type the veins are irregularly spread in a reticulate fashion (Jālavat, cf. Reticulate venation). The leaves with parallel venation are called Mauñjaparṇa and those with reticulate venation are termed Jālikāparṇa. [Verse 25].

Patra pakṣaṁ (Leaf lamina)

The Patrapakṣaṁ or the lamina is of four types, viz. Samapakṣa, Viṣamapakṣa, Samakarṇika and Viṣamakarṇika.

In Sampakṣa type the two lateral parts are symmetrical while in Viṣamapakṣa type they are asymmetrical. In

Samakarṇika type the incisions on the leaf blade are regular while in Viṣamakarṇika type they are irregular. [Verse 26].

Rasa Koṣa (Cells)

The leaf cells are the reservoirs of the fluid (Rasa). [Verse 27]. The cells' are also called Aśaya which means a receptacle. A leaf is made up of innumerable cells, each being surrounded by a membrane (Kalā), containing the fluid constituted of the five elements of the Pāñcabhautic guṇa. Cells are microsocopic (Aṇavaśca) and contain a kind of coloured substances (Rañjaka). The membrane of the cell is fine and transparent. It is derived out of the Kalalaṁ (jelly like substance) as a result of the metabolic changes under the influence of the heat energy (Bhūtoṣmā). [Verse 28].

Mārhi (Rachis)

It is the axis which bears the petioles in a compound leaf as found in the leguminous plant (Simbiparṇī). It falls off when the leaves are old. It may be of two types, viz. Ekasandhikaṁ and Bahusandhikaṁ. In Ekasandhikaṁ type all the leaflets are articulated together at a point (cf. Palmạte compound). In Bahusandhikaṁ type the leaflets in pairs are successively borne along the two sides of the rachis (cf. Pinnate compound).

Vistara and Vitana (Tendrils)

In some creepers there is an additional organ like a rachis; it remains attached to the stem permanently like a branch. It is called Vistāra (cf. Tendrils). In some cases, the terminal pair of leaves (leaf-lets) appear like threads (cf. Tendril of *Antigonon*). In some creepers, a thread like outgrowth comes out from the base of the petiole. It is termed Vitāna (cf. Axillary stem tendril of Passion flower). [Verse 29].

Paṭṭikaṁ (Sheathing leaf base)

The Paṭṭikaṁ or the sheathing leaf base is found in the plants belonging to the grass family. It is attached to the leaf. It ensheathes the stem at the nodes and at the apex. It falls off along with the old leaf. It is also called 'Khollaka'. [Verse 30].

Śungaṁ (Bud scale)

Śungaṁ is the organ present in some plants which covers the leaf bud (Nava pallava) as well as the flower bud (Koraka). [Verse 31]. (This particualr organ is present in plants like *Ficus religiosa* and *F. bengalensis.* In Pārāśarian terminology. *F. bengalensis* is temed 'Śungī' as it bears the organ śungaṁ).

Upakaṣaṁ (Stipule)

In some plants the petiole bears an outgrowth on each side. It is called Upapakṣam (cf. Stipule). [Verse 32].

Sāmanyatah patraṁ (Variation of lamina)

The lamina may be of two types :

Samaparnaṁ and Viṣamaparṇaṁ. A leaf is termed Samaparṇaṁ when its two lateral parts are regular; and it is Viṣamaparṇaṁ when they are irregular. [Verse 33].

Nāmatah Patra-Gaṇa-Samgraha

(The Nomenclature of different types of leaves):

Arghya patra*, Juhū patra, Sruva patra, Vṛttavarha, Maṇḍalāgra, Dīrghadala, Vīra patra, Vāṇavarha, Śankuvarha, Sūci patra, Īliparṇī, Karṇikachada, Triparṇa, Krakacchada, Vṛjina patra, Guccha patra, Śrīparṇī, Simbiparṇī, Yugma patra, Nakha patra, Bāla patra,

*. In Coochehar (West Bengal) the above plant is called 'Arghya pattia'.

Karachada, Mañujabarha, Cīri patra, Kārabīdala, Kaca Patra, Sūrpaparṇa, Viṣama-pakṣaparṇa, Bali Patra, Citraparṇa, Pṛśniparṇī, Maṇḍukaparṇī, Śuktiparṇī, Kamaṭhachada, Godhāpadī, Gokarṇī, Aśwakarṇī, Hastikarṇī, Hastiparṇī, Hamsapadī, Pārāvatpadī, Lomaśaparṇī, Kaṇṭaparṇī, Nādidala, Pīvarīdala, Mauñja patra. [Verse 34-35].

The above different types of leaves will now be described mentioning their characteristic features. [Verse 36].

Arghya patraṁ - Leaves in some trees, shrubs and creepers are shaped like an 'Arghya-pātra' a receptacle used for offering flowers etc. in a religious rite. Here the attachment of the petiole with the lamina is of Pṛṣṭhagranthika type (cf. peltate e.g. *Stephania hernandifolia*). [Verse 37].* Fig. 2a.

Juhū patraṁ - In many trees, shrubs and creepers, the leaves are shaped like a 'Juhū pātra' - a kind of ladle used to offer butter oil to the sacred fire as in ancient Vedic rites. Here the petiole remains attached to the base of the lamina - Prāntagranthika (e.g., *Ficus bengalensis*). [Verse 38].

Sruva patraṁ - The leaves are shaped like a 'Sruva' which is a Vedic sacrificial implement. Here the petioles are attached to the base of the lamina. [Verse 39].

Vṛttavarha - Here the leaves are circular in shape with a peltate petiole (cf. Rotund e.g. *Nelumbo nucifera*). [Verse 40].

Maṇḍalāgra - A leaf which has a round apex is called Maṇḍalāgra, (e.g. *Artocarpus heterophyllus*). The petiole is attached to the base of the lamina. [Verse 41].

Dīrghadala - A leaf which has long lamina with the petiole being inserted at its base is called Dīrghadala (e.g. *Musa paradisiaca*). [Verse 42].

Vīrapatra - The leaf is shaped like a sword. The lamina at its base is attached with a bulb below (cf. Lanceolate e.g. *Lilium* sp.). [Verse 43]. Fig. 2b.

Vāṇavarha - A leaf which is triangular in shape, resembling an arrow head (Vāṇaphalaka) is called Vāṇavarha (cf. Hastate e.g. *Ipomoea aquatica*). [Verse 44]. Fig. 2c.

Śankuvarha - A leaf which is shaped like a spear is called Śankuvarha. [Verse 45].

Sūciparṇī - A leaf which is shaped like a needle is called Sūciparṇī (cf. Acicular, e.g. *Pinus* sp.). [Verse 46].

Īliparṇī - A leaf which is shaped like a knife is called Īliparṇī. [Verse 47]. Fig. 2d.

Karṇikāchada - A leaf with many angular lobes is called Karṇikāchada e.g. *Cucurbita* sp.

Triparṇa - A compound leaf with three leaflets borne on a common petiole is called Triparṇa (cf. Trifoliate or Ternate). The lateral veins in the leaflets extend transversely (e.g. *Aegle marmelos*). Triparṇa is often termed as Gucchaparṇa. [Verse 48].

Krakacchada - A leaf is called Krakacchada when its margin is serrated like the edge of a saw (cf. serrate e.g. Pandanus sp). [Verse 49]. Fig. 2e.

Vṛjina patra - A leaf which appears wavy and curled is called Vṛjina patra (cf. Repand e.g. *Polyalthia longifolia*). [Verse 50]. Fig. 2f.

Guccha patra - A compound leaf with many leaflets is called Gucchapaṭra. They remain attached to the petiole at the base [Verse 51].

Śrīparṇa - A leaf almost triangular (Tryaśri) in shape with a pointed apex and round base is called Sriparṇa (cf. Ovate, e.g., *Hibiscus rosa-sinensis*). [Verse 52].

Simbiparṇa - A compound leaf with three leaflets which appears like a bird (with its spreadng wings). Of the three leaflets, the middle one is Śrīparṇa type (cf. Ovate). The two lateral ones appear like the wings of a bird. The petiole is partially bent. Often many pairs of leaflets remain arranged along the rachis. (cf. Pinnate compound e.g. *Tamarindus indica*). [Verse 53].

Yugma patra - A leaf is called Yugma patra when its lamina is deeply notched both at the base and at the apex as if two leaves are merged together sidewise. The petiole is articulated at the notch of the base (Prāntagranthika cf. Obcordate e.g. *Bauhinia* sp.). [Verse 54].

Bālapatra - Leaves which are very small in size, having a round base and apex are called Bālapatra. (e.g. *Trianthema portulacastrum*). [Verse 55].

Nakhapatra - A leaf which is shaped like a nail is called Nakhapatra (Lunate, e.g. *Passiflora lunata*). [Verse 56].

Karachada - A leaf which looks like a palm is called Karachada (cf. Palmate compound). [Verse 57].

Mañjuvarha - A leaf which is shaped like an eye (Mañju means eye) is called Mañjuvarha. The petiole remains attached to the base of the lamina. (e.g. *Cinnamomum* sp.) [Verse 58]. Fig. 2g.

Cīripatra - A leaf which appears like a membrane (Jhillikāya sadṛśaṁ) with an irregularly incised lamina (Viṣamakarṇika) is called Cīripatra (cf. a Lyrate leaf e.g. *Raphanus sativus*). Here the petiole is attached to the base of the lamina. [Verse 59].

Kāravīdala - A leaf with highly incised lamina resembling the stamens of a flower is called Kāravīdala (cf. Decompound leaf, e.g. *Foeniculum vulgare*). [Verse 60].

Kacapatra - A leaf (with a very finely divided lamina) appearing like a tuft of hairs is called Kacapatra. [e.g. *Asparagus racemosus*). [Verse 61].

Sūrpa patra - A leaf with a large lamina looking like a winnowing fan (Sūrpa) is called Sūrpapatra. (e.g. *Alocasia* sp.). [Verse 62].

Viṣamapakṣaparṇa - A leaf whose lamina is asymmetrical is called Viṣamapakṣaparṇa. (e.g. *Pterospermum suberifolium*). [Verse 63].

Baliparṇa - A leaf whose surface is rough like an elephant's skin is called Baliparṇa. It is also called Kharapatra. (cf. Scabrous leaf, e.g. *Ficus cunia*). [Verse 64].

Citraparṇa - A leaf whose surface is adorned with colourful spots and lines is called Citraparṇa. (e.g. Croton). [Verse 65]. Fig. 2h.

Pṛśniparṇa - A small round leaf which has a thin lamiṇa with rough surface is called Pṛśniparṇa. The petiole is articulated to the base of the lamina. (e.g. *Uraria lagopoides*). [Verse 66].

Maṇḍukaparṇa - A leaf whoṣe overall shape gives the impression of a frog (cf. Reniform leaf, e.g. *Centella asiatica*). [Verse 67]. Fig. 2i.

Śuktiparṇa - A leaf is called Śuktiparṇa when its lamina is shaped like a mother of pearl (cf. an elliptical leaf, e.g. *Psidium guajava*). [Verse 68]. Fig. 2j.

Kamaṭhaparṇa - A leaf which assumes the outline shape of a tortoise is called Kamaṭhaparṇa (cf. Rotund, e.g. *Nelumbium* sp.). [Verse 69].

Godhāpadī patra - A leaf whose shape resembles the foot of an Iguana is called Godhāpadī patra, (e.g. *Cayratia pedata*). [Verse 70].

Gokarṇa - A leaf which is shaped like the ear of a cow is called Gokarṇa. [Verse 71].

Aśwakarṇa - A leaf which is shaped like the ear of a horse is called Aśwakarṇa. (e.g. *Shorea robusta*). [Verse 72].

Hastikarṇa - A large leaf whose shape is like that of an elephant's ear is called Hastikarṇa. (e.g. *Alocasia macrorrhiza*). [Verse 73].

Hastiparṇī - A leaf whose shape resembles that of an elephant's head along with its dropping trunk is called Hastiparṇī. It is also called Pīluparṇī. (e.g. *Abroma augusta*). [Verse 74]. Fig. 2k.

Hamsapadī - A leaf which is shaped like a swan's foot is called Hamaspadī. [Verse 75].

Pārāvatpadī - A leaf which is shaped like the foot of a pigeon is called Pārāvatpadī (cf. Palmately lobed leaf, Pedate type, e.g. *Vitis pedata*). [Verse 76].

Lomaśaparṇa A leaf whose lamina is covered with hairs is called Lomaśaparṇa. [Verse 77].

Kaṇṭaparna - A leaf is called kaṇṭaparṇa when its surface is covered with spines and hence difficult to touch (dusparśa cf. Spinose, e.g. *Solanum melongena*). [Verse 78].

Nādiparṇa - A leaf is called Nādiparṇa when its lamina is tubular in form (Suṣira - e.g. a kind of aquatic plant used for mat making in Bengal). [Verse 79].

Pīvarachada - A leaf which is thick, fleshy and juicy (sāndra rasa) is termed Pīvarachada (cf. Succulent, e.g. *Bryophyllum, Aloe vera*) [Verse 80].

Mauñjaparṇa - The leaves of the plants belonging to the grass family (Tṛnavallaja) having characteristic parallel venation are called Mauñjaparṇa. [Verse. 81].

Puṣpapatraṁ - It is a special type of leaf which appears immediately before flowering as a forerunner. (cf. Bracts, e.g. Bracts of *Hibiscus rosa-sinensis,* and *Poinsettia pulcherrima*). [Verse 82]. Fig. 2l.

Out of the innumerable trees, creepers and shrubs only those leaves that have been found, so far, have been described. Those which have not been mentioned may be characterised, similarly, according to their morphological features. [Verse 83].

Thus ends the fourth chapter of the Bījotpattikāṇḍa of the Vṛkśāyurveda narrated by Parāśara. [Verse 84].

बीजोत्पत्तिकाण्डः
पुष्पाङ्गसूत्रीयाध्यायः ।

अथातः पुष्पाङ्गसूत्रीयमध्यायं व्याख्यास्याम इत्याह पराशरः ।।१ ।।

खलू पुष्पन्तु फलव्यञ्जनं कुसुमं सुमनः प्रसूनमित्यनर्थान्तरम् ।।२ ।।

पुष्पाणामभिव्यञ्जनं पुष्पपत्रं खलु प्राक्पुष्पमेव खुड्डुकपत्रञ्चेति ।।३ ।।

अथ कुट्मलः पुष्पकलिका कोरकश्चेति ।।४ ।।

इह खलु वल्लरिवृन्तजालकदलकेशरपरागकिञ्जल्कपरिमलमकरन्द बीजाधारवराटकस्थालकानीति पुष्पाङ्गानि भवन्ति ।।५ ।।

अथ पुष्पाणां सामान्यलक्षणमुपदेक्ष्यते ।।६ ।। तत्र कोरकस्यावरणं जालकम्। तदभ्यन्तरतः पुष्पदलम्। ततः केशरश्च। ततो बीजाधारस्तत्रोपरि केशरवेष्टितैर्वराटकश्च तत्र बीजाधानं भवति। परागस्तु रजश्च केशराणाम्। किञ्जलकं परागाधार एव केशराणां शीर्षे सन्धितम्। तदेव प्रकृत्या द्विविधम् । प्रान्तसन्धितं पृष्ठसन्धितञ्च । परिमलः पुष्पगन्धः स्नेहश्च । पुष्पमधु मकरन्दश्च तस्याधारः पुष्पोदरञ्च ।।७ ।।

अथ खलु वल्लरिस्तु शखाग्रे वा वर्तिता पुष्पवृन्तानामभिवाहिका मञ्जरिश्च। अष्टविधा वल्लरि र्भवति स्वरूपतश्चतद् यथा पलाशवल्लरिपङ्क्तिकारकाछत्रागुच्छवल्लरिसङ्कुलौतुपुच्छिकाक्षमञ्जरिकाश्चेति ।।८ ।। तत्र पलाशवल्लरिस्तु पुष्पपत्रेण युक्ता भवति ।।९ ।। पङ्क्तिमञ्जरिकायां पङ्क्तिक्रमेण पुष्पबन्धनं विद्यात् ।।१० ।। अरा इव चक्रनाभौ यथा

संहतास्तथभूता पुष्पबन्धनादरकवल्लरिः इति संज्ञायते ।।११।। छत्रा तु छत्रस्य कारवसंस्थानैश्च कारविका विज्ञायते ।।१२।। गुच्छवल्लरिस्तु गुच्छेन विद्यात् ।।१३।। संकुलायां मिश्रेण वैषम्येण वा पुष्पबन्धनं विद्यात् ।।१४।। ओतुपुच्छिकातु मार्ज्जारपुच्छसंस्थाना ।।१५।। अक्षमञ्जरिका त्वक्षकाण्डमाचक्षते तद्व्यासतः काण्डगणीये वक्ष्यते ।।१६।।

पुनश्च वल्लरिर्द्विविधं भवति सशाखाशास्वाभ्याम् । तत्र सशाखवल्लरिस्तु शाखया विद्यते। अशाखवल्लर्यां न च शाखा दृश्यते ।।१७।। वृन्तं पूर्वमुद्दिष्टं ।।१८।।

अथ पुष्पमण्डलम् । पुष्पवृन्ते जालकदलकेशराणां सन्निवेशमण्डलं त्रिविधं भवति। यथा वृत्तमणड्लं पृथग्मण्डलं मिश्रमण्डलञ्चेति। तत्र वृत्तेन यद्दलकेशर बन्धनं तद्वृत्तमण्डलम्। पङ्क्त्या यत् सन्निवेशं तत् पृथग्मण्डलम। मिश्रेण मिश्रमण्डलम् ।।१९।। पुष्पाणां जालकदलकेशराश्च प्रायशो वृत्तमण्डलेन वृन्ते सन्निविष्टं भवति कदाचिद् वैषम्येण विद्यात् ।।२०।।

इ॰ खलु जालकं क्षारकं वप्रमित्यनर्थान्तरम् ।।२१।। प्रकृत्या जालकं हि द्विविधं भवति। पुष्पान्त जालकं स्थिरजालकञ्च। तत्र पुष्पान्त जालकं फलाभिव्यक्ते शुष्कं प्रपतितं भवति। स्थिरजालकन्तु शुष्कं न जायते फलाभिव्यक्तेऽपि। कस्यचित् स्थिरजालकं वृन्तेन सह संहतं तच्च वृन्ताकजालकमाचक्षते एतदेव यद् यद् पुष्पफले दृश्यते तद् तद् पुष्पफलं वृन्ताकमाचक्षते ।।२२।। कस्यचित् स्थिरजालकं स्थालकेन सह संहतञ्च। तत्र जालकं स्वातन्त्रयेण न विद्यते। एतदेव पिण्डपुष्पेषु दृश्यते ।।२३।।

जालकं पुनर्द्विविधं। युक्तजालकं मुक्तजालकञ्च। तत्र युक्तजालकं कर्णेन परस्परं युक्तं भवति। मुक्तजालकन्तु विपर्ययेण विद्यात् ।।२४।।

आकृतितो जालकं पुष्पदलेन सह प्रायशः साम्यं भवति । कदाचिद् विशेषेण विद्यात् । तत्र कस्यचित् पुष्पगणस्य जालकं समदलं समकर्णितं भवति। कस्यचिद् गणस्य विषमदलं विकर्णितञ्च वा। कस्यचिज्जालकमजिनवत् पुष्पस्य छादकं भवत्येवमजिनजालकमाचक्षते। कस्यचित् गणस्य जालकं जृम्भितं तच्च जृम्भजालकमाचक्षते। कस्यचिज्जालकं युक्तंकर्णं वा भवति ।।२५।।

कस्मिश्चिद् गणे वृन्ते वा जालकस्य निम्ने क्षुल्लकपत्रसंङ्काशमुपजालकं दृश्यते ।।२६।। तृणवल्वजानामेव कण्डूलानामक्षमञ्जरिकायां पुष्पकोरकानामाच्छादकमेककर्ण पट्टिकसङ्काशमधिकमङ्गं दृश्यते तच्च मोचिकमाचक्षते ।।२७।।

यथा जालकं पुष्पकोरकस्य छादकं भवति तथा पुष्पदलञ्च केशरवराटका नामाच्छादकञ्च ।।२८।। प्रकृत्या चतुर्विधं हि दलं भवति। तद् यथा मुक्तदलञ्च युक्तदलञ्च केशराङ्कश्च स्वैरञ्चेति ।।२९।। तत्र मुक्तदलं विभक्तमसंहतदलेन विद्यात । युक्तदलन्तु सङ्घातेन विद्यादेवं कर्णेन युक्तं भवति ।।३०।। केशराङ्कश्च केशरेण सह संहतंभवति। स्वैरदलञ्च स्वातंत्र्येण विद्यात् । एतदेवं तुङ्गपुष्पमण्डले जालककर्णान्तरे संविशति ।।३१।। कस्यचिद् गणस्य पुष्पं समदलं समकर्णञ्च भवति। कस्यचिद् विषमदलं विकर्णञ्च वा ।।३२।।

आकृतितो मुक्तपुष्पदलं प्रायशः पर्णसङ्काशं भवति। कस्यचिद् पुष्पस्य दलानि केशलोमशूकसदृशनिभवन्ति । तदेव लोमशपुष्पं शूकपुष्पं वा प्रचक्षते ।।३३।। कस्यचिद युक्तदलं घण्टासंस्थानम् । कस्यचिद् घटसङ्काशम् । कस्यचिद नलकाकारमेव शीर्षकर्णितम् । कस्यचित् सिंहमुखेन सदृशं जृम्भितञ्च। कस्यचित् कपोतवक्त्रसङ्काशम् । कस्यचिद् नासापुटसङ्काशं युक्तदलञ्च भवति। युक्तदलपुष्पस्य यदङ्गं नलकाकारं तच्च कुण्डसंज्ञकमेव कुण्डमण्डलं विज्ञायते ।।३४।।

केशरास्तु किञ्जल्काधाराः । तेषामधिष्ठानभेदेन त्रिधा भिद्यते । यथा दलोत्सङ्गश्च स्थालकीश्च वराटोत्सङ्गश्चेति ।। तत्र दलोत्सङ्गकेशरस्तु दलेन सह संसक्तं भवति। स्थालकीस्तु स्थालकात् सञ्जायते एव मुक्तकेशर इत्याचक्षते। वराटोन्सङ्गश्च वराटे नलकसदृशेन संघातेन संवीतं भवति। एव संहतकेशरश्च । आयामभेदेन पुनर्द्विविधः समकेशरो विषम केशरश्च। तत्र समकेशरो दैर्ध्येण समो भवति विषमस्तु वैष्यमेण विद्यते ।।३५।।

आकृतितो नामतश्च केशराः पुनः पञ्चविधा भवन्ति तद् यथा तुङ्गश्च मञ्जुश्च पन्नगीश्च वाजीक्रान्तश्च उर्मिकेशरश्चेति। तत्र तुङ्गस्तु तुङ्गशीर्षेण किञ्जल्केन

विद्यात् । मञ्जुश्च मञ्जुलेन किञ्जल्केन विद्यात् । पन्नगीश्च पन्नगफण इव किञ्जल्केन विद्यात् । वाजीक्रान्तस्य वाजीपदलाञ्छनं किञ्जलकं भवति । उर्मिकेशस्तुर्मिवत् वराटकाच्छादकश्चैव कर्णसंस्थानेन परस्परेण संहतश्च ।।३६।।

अथ बीजाधारः फलसंज्ञकश्च खलु बीजानामाधार आशयश्च। सतु सन्धितासन्धिताभ्यां द्विविधो भवति। तत्र सन्धितबीजाधारस्तु विदरसंज्ञकं भवति। असन्धितस्तु कुड्यसंज्ञकमाचक्षते ।।३७।।

ततो बीजाधारश्च खलु प्रकृत्या कश्चिदेकवर्तको बहुवर्तको वा कश्चिद भवति। वर्तकस्तु पुप्लिकेन च पुषेण वा विभक्त आशयः पुटकश्च । पुनरेव बीजाधारस्त्रिविधं भवति पुष्पभेदेन यथा पुष्पक्रान्तश्च संवृतश्च पुष्पशीर्षकश्चेति। तत्र पुष्पक्रान्त बीजाधारः पुष्पगर्भे सन्निविष्टं भवति । संवृतबीजाधारस्तु यस्मिन्नुलुखलनिभदलकेशरसङ्घाताङ्गेन संवृतं भवति। पुष्पशीर्षकबीजाधारश्च बीजाधारशीर्षे पुष्पं सन्निविष्टं भवति ।।३८।।

पुष्पगर्भे केशरवेष्टिते बीजाधारशीर्षे यस्मिन्नङ्गे बीजाधानं भवति स तु वराटकः। स च पुष्पगणविशेषेण ह्येकशीर्षं द्विशीर्षं बहुशीर्षञ्च भवति। एव युक्तशीर्षं वा मुक्तशीर्षञ्च। कस्यचिद् गणस्य वराटको दीर्धं नलकाकारञ्च क्वचिद् खर्वं कश्चिद् वर्तुलं वा भवति। ततश्च वराटबीजाधारयोः सन्धिर्वॅराटसङ्गमश्च ।।३९।।

यदा वराटके परागो मकरन्देन सह सम्प्रयोगं भवति तदा बीजकोषे बीजमावर्तते कालप्रकर्षाच्च ।।४०।।

अथ खलु पुष्पाधारस्तु स्थालकं यत्तु कस्यचिद् गणीयस्य पुष्पवृन्तस्योपरि संहतं भवति । कस्यचिद् गणस्य जालकस्योपरि स्थालकं दृश्यते। प्रकृत्या स्थालकं त्रिविधम् । तद यथा कुण्डं फल्गु पिण्डञ्चेति । तत्राद्यं शरावसंस्थानम्। द्वितीयमन्तरावकाशं वृत्तपरिमण्डलञ्च। तृतीयं पिण्डसंस्थानमायतं वर्तुलं वा भवति ।।४१।।

खलु पुष्पाणामधिष्ठानतः प्रकृतितश्चाकृतितश्च नामतश्च वर्णतश्च गणविभागं प्रत्यक्षेण यदुपलभ्यते तदनुव्याक्ष्यास्यामः ।।४२।। तत्राधिष्ठानभेदेन पुष्पन्तु

त्रिविधं विद्यात् । यथा कुक्षिपुष्पञ्च काण्डपुष्पञ्च वल्लरिपुष्पञ्चेति ।।४३।।

यानि पुष्पाणि खल्वेकैकेन वा गुच्छेन पत्रकुक्षौ सन्निविष्टानि तानि कुक्षिपुष्पगणीयानि भवन्ति । एतदेव त्वेकपुष्पवल्लरिः केचिदाचक्षते ।।४४।।

यानि पुष्पाणि एकैकेन वा गुच्छेन काण्डे समुद्भवन्ति तानि काण्डपुष्पगणीयानि भवन्ति ।।४५।।

यानि पुष्पाणि शाखाग्रे वल्लयां सन्निविष्टानि तानि वल्लरिपुष्पगणीयनि स्युः ।।४६।। केषाञ्चित वल्लरिपुष्पाणां वृत्तं नास्ति तानि पुष्पाणि बहुनि खल्वेकजालकेन वेष्टितानि वल्लयाँ कुण्डस्थालकस्योपरि कुर्चाकारेण सन्निविष्टानि भवन्ति एतानि वल्लरिपुष्पाणि कुचसंज्ञकानि। तेषां वीजाधारस्तु पुष्पशीर्षकञ्च भवति। एतदेव कुर्चपुष्पं संज्ञायते ।।४७।।

पुष्पञ्च प्रकृत्यैव पुनः सफलनिष्फलभेदेन द्विविधं विद्यात्। तत्र बीजकोषयुक्तत्वेन सफलं भवति। निष्फलन्तु विपर्ययेण। तस्मात् केचित् वृक्षाः सफलाः केचित् निष्फलाश्च भवन्ति। कदाचिदेकस्मिन् वृक्षे युगपत् सफलं निष्फलं पुष्पञ्च दृश्यते। एवञ्च वल्लर्याम् ।।४८।।

प्रकृत्या पुनः कश्चित् पुष्पगण एक बीजाधारः कश्चिद् द्विबीजाधारो बहुवीजाधारश्च कश्चिद् भवति ।।४९।।

केषाञ्चित् पुष्पगणानां कुट्मलो वल्लयाँमूर्ध्वमुखञ्च विकशिते विपर्ययं भवति । एष गण प्रत्यक्पुष्पं विज्ञायते ।।५०।।

पुनः पुष्पं व्यक्ताव्यक्तकेशरभेदेन द्विविधम। तत्र व्यक्तकेशरपुष्पाणां केशरश्च दलेन सह व्यक्तो भवति। अव्यक्तकेशराणान्तु केशरश्च कुण्डेन संवृतत्वाद् व्यक्तो न भवति ।।५१।।

पुष्पं पुनरेव, द्विविधं विद्यात् सोमसूर्यसम्बन्धात् । तत्र पुष्पाणि यानि विकशितानि रजन्यां तानि चन्द्रकान्तगणीयानि भवन्ति निशा हि चन्द्रबलवत्त्वात् । यानि तु पुष्पाणि दिवायां विकसन्ति तानि रविकान्तगणीयानि दिवसं हि रविबलत्त्वात् ।।५२।।

ऋतुपुष्पसदापुष्पभेदेन पुनरेव द्विविधं। तत्र निर्दिष्टकालजातानि पुष्पाणि ऋतुसंज्ञकानि । अनिर्दिष्टकालजातानि तु सदापुष्पणि स्युः ।।५३।।

द्विविधं हि पुनः पुष्पं व्यक्ताव्यक्तभेदेन । तत्र व्यक्तपुष्पाणां दलकेशरवराटकानि स्फुटानि स्युः । अव्यक्तानान्तु दलकेशरवराटकानि फल्गुस्थालकेन संवृतत्वाद् न दृश्यन्ते ।।५४।।

पुनः पुष्पं दलभेदेन द्विविधं। मुक्तदलपुष्पं युक्तदलपुष्पञ्च। तत्र मुक्तदलपुष्पं स्वतंत्रदलेन विद्यात्। युक्तदलपुष्पन्तु संहतदलेन भवति ।।५५।।

पुनरेवपुष्पं द्विविधं संस्पृष्टसंस्फुटाभ्याञ्च । तत्र संस्पृष्टदलपुष्पं यच्च दलेन परस्परेण श्लिष्टं भवत्येवं दलानां कृत्स्नाङ्गमस्फुटञ्च। संस्फुटदलन्तु परस्परेण न च श्लिष्टं भवत्येवं स्फुटतरञ्च ।।५६।।

अङ्गभेदेन पुनरेव पुष्पं द्विविधम् । समाङ्गपुष्पं हीनाङ्गपुष्पं च । तत्र समाङ्गपुष्पं दलकेशरवराटकबीजाधारैश्च कृत्स्नाङ्गं भवति । हीनाङ्गपुष्पन्तु विपर्येण एवं द्विविधेन भिद्यते। आद्यं बीजाधारहीनेन मञ्जुपुष्पमाचक्षते केशरपुष्पञ्च। द्वितीयं दलकेशरहीनेन रण्डपुष्पं विज्ञायते श्रीपुष्पञ्च । बीजाधारहीनत्वाद् मञ्जुपुष्पं निस्फलं च भवति । कदाचिदेकस्मिन् वृक्षे वल्लर्याञ्च युगपदुभयं पुष्पं दृश्यते । रण्डपुष्पन्तु प्रायशस्त्रिवर्तकं त्रिशीर्षवराटकं भवति । यदैकस्मिन् वल्लयाँ युगपदुभयपुष्पं भवति तदा स गणो मञ्जुश्रीपुष्पेण संज्ञायते । कस्यचिद् गणस्य श्रीपुष्पस्य दलभस्ति ।।५७।।

अथ बीजाधारव्यञ्जनेन पुष्पमण्डलं चतुर्विधं भवति । तद् यथा तुन्दमण्डलं कुम्भमण्डलं तुङ्गमण्डलं वाटयमण्डलं चेति । तत्र तुन्दमण्डलगणीयस्य दलकेशराश्च बीजाधारनिम्ने सन्निविष्टा भवन्ति तस्मात पुष्पक्रान्तबीजाधारो भवति। कुम्भमण्डलगणीयस्य दलकेशरास्तु बीजाधारशीर्षे सन्निविष्टा भवन्ति। तस्मात् पुष्पशीर्षबीजाधारो भवति। एतदेव युक्तमुक्तदलाभ्यान्तु द्विविधम्। तुङ्गमण्डलगणीयस्य कुण्डस्यालकस्योपरि जालकदलोन्सङ्गकेशरैश्च वृत्तमण्डलेन वराटं संवेष्टयेत् बीजाधारश्च स्थालकोत्सङ्ग भवति। वाटयमण्डलगणीयस्य बीजाधारवराटकयोः सङ्गमे तूलुखल सदृशसंहतदलकेशैश्च नलकाकारेण बीजाधारमावृतं भवति। तस्माद् वराटोत्सङ्गकेशराश्च भवन्ति चैव संवृतबीजाधारश्चेति ।।५८।।

खलु पुष्पाणामाकृतितश्च क्वचित् समगणीयं क्वचिद् विषमगणीयञ्च भवति । तत्र समगणीयन्तु दल केशरेषु समसंख्येयं भवति। विषमगणीयञ्च वैषम्येण विद्यात् । यदैकस्य जात्याः पुष्पलक्षणे भिन्नजात्याः फललक्षणं लक्ष्यते तदा तत् पुष्पं वञ्जुलं संज्ञायते ।।५९।।

पुष्पाणि प्रायशः पञ्चदलानि पञ्चकर्णानि भवन्ति । यान्युपर्युपरि निसृष्टानि बहुदलानि तानि स्ववकपुष्पाणि स्युः ।।६०।।।

स्तवकपुष्पानामन्तरमण्डल बहिर्मण्डलाभ्यां संज्ञायते ।।६१।।

खलु वर्णतः पुष्पाणि नानाविधानि स्युः । तद् यथा श्वेतपीतरक्तारुणश्यावनीलविविधानि भवन्ति। एतैश्च वर्णैश्च वृक्षवल्लीगुल्मानां वर्णो बुभुत्सितव्य इति ।।६२।।

अपरिसंख्येयानां वृक्षवल्लीगुल्मानांपुष्पानां प्रकृतिनामलिङ्गानि ह्यसंख्यानि भवन्ति । अत्रैव नामतः कतिपयाविष्कृततमा मण्डलोक्तपुष्पगणा उदाहृताः स्वभावाश्च। इह ये चानुक्ता गणास्ते स्वे स्वे काण्डे वक्ष्यन्ते ।।६३।।

चूतगणीयम्

अथातः संप्रवक्ष्यामि चूतपुष्पस्य लक्षणम् ।
मण्डलं चूतपुष्पस्य विमुक्तं दलकेशरम् ।।
खुड्डाकं तुन्दमण्डलंसशाखमधुमञ्जरिः ।
जालकञ्च दलञ्चैव पञ्चभवति तत्र वै ।।
आरक्तखर्वकिञ्जल्कैः पञ्चैव केशरान्वितम् ।
वराटैकं भवेत्तत्र मुक्तश्चैव समुच्छ्रितम् ।।
पुष्पक्रान्तं फलं तत्र कुड्याख्यमेकवर्तकम् ।
बीजमेकं भवेत्तत्र कीरवोसंकाष्टिकं स्मृतम् ।।
बीजस्य वन्धनं तत्र वृन्तान्तिकेन दृश्यते ।।६४।।

समीगणीयम्

समी तु तुन्दमण्डला विषमा विदला स्मृता ।
पञ्चमुक्तदलैश्चैव युक्तजालककर्णितैः ।।
दशभिः केशरैर्विद्यात् समीपुष्पस्य लक्षणम् ।
समी सिम्बिफला ज्ञेया पार्श्वबीजा भवेच्च सा ।।
वक्रं विकर्णिकपुष्पं शूकाख्यपुष्पमेव च ।
एतैश्च पुष्पभेदैस्तु भिदयन्ते समिंजातयः ।।६५।।

पुप्लिकगणीयम्

पुप्लिकपुष्पगणीये मुक्तञ्च दलकेशरम्
विषमं तुन्दमण्डलं मञ्जरिमाधवोद्भवा ।।
जालकं पञ्चविज्ञेयं दलञ्चैव तथा भवेत् ।
केशरैर्बहुभिश्चैव किञ्जल्कं प्रान्तग्रन्थिकम् ।।
विभक्तं वराटञ्चैव शीर्षन्तु बहुकर्णितम् ।
अखण्डं फलवल्कलं शस्यञ्च केशरनिभम् ।।
पुष्पक्रान्तं फलं शस्यं विभक्तं बहुपुप्लिकैः ।
दृश्यन्ते बहुबीजानि पुप्लिकान्तर्गतेन च ।
एतानि लक्षणानि हि भवन्ति पुष्पमण्डले ।।६६।।

स्वस्तिकगणीयम्

अथातः स्वस्तिकपुष्पं वक्ष्यामि सर्व लक्षणैः ।
स्वस्विकं तुन्दमण्डलं वल्लर्यां पङ्क्तिबन्धनम् ।।
जालकं स्वस्तिकाकारं पुष्पान्तं मुक्तमेव च ।
चतुर्दलञ्च मुक्तञ्च षड्मुक्तकेशरान्वितम् ।।
तेषां द्वौ खर्वकेशरौ भवतः पुष्पमण्डले ।
युग्मं वराटकं युक्तं फलञ्च द्विपुटं भवेत् ।।
सदृशं समिफलेन सन्धितं फलवल्कलम् ।
एतैश्च लक्षणैर्विद्यात् स्वस्तिकपुष्पमण्डलम् ।।६७।।

त्रिपुषगणीयम्

ततश्च त्रिपुषपुष्पं वक्ष्यामि पूर्णलक्षणै: ।
त्रिपुषं कुम्भमण्डलं निष्फलञ्चापि दृश्यते ।।
प्रायश: कुक्षिपुष्पञ्च मिश्रवल्लरिका क्वचित् ।
जालकञ्च दलञ्चैव युकञ्च पञ्चकर्णितम् ।।
त्रिशीर्षैस्त्रिभि: केशरैस्त्रिशीर्षवराटेन च ।
त्रिभि: पुषैस्त्रिवर्तकै: पुष्पञ्चास्य विशिष्यते ।
बहुलपृथुक बीजं दृश्यते त्रिपुषफले ।।६८।।

अवाकपुष्पगणीयम

अवाक्पुष्पं भवेत्तत्र यत्र हि छत्रमञ्जरि: ।
छत्रस्य कारवाकारा तस्मात् कारविका स्मृता ।।
सूक्ष्ममुक्तदलं पञ्च सूक्ष्मकेशरकन्तथा ।
द्विपुटं कुम्भमण्डलं द्विवराटं द्विवर्तकम् ।।
बीजकबीजाधारौ तु भवत: पुष्पमण्डले ।।६९।।

अक्षपुष्पगणीयम्

अक्षपुष्पं प्रवक्ष्यामि शृणु मे द्विजसत्तम ।
सशाखमञ्जरिपुष्पं खुल्लकं पुष्पमण्डलम् ।।
कुम्भसंज्ञं भवेत्तत्र पञ्चकर्णैर्वा जालकम् ।
पञ्चभिश्चैव दलैर्वा केशरैर्दशभिस्तथा ।।
वराटैकं भवेत्तत्र फलञ्च कुड्यसंज्ञकम् ।
अस्त्रकफलबीजञ्च कीखोसं कर्परं भवेत् ।।
तत्रैकं बीजयुक्तञ्च वृन्तान्तिकेन दृश्यन्ते ।।७०।।

कुहपुष्पगणीयम्

कुहपुष्पं प्रवक्ष्यामि तुङ्गेन यच्च भाषितम् ।
सफलं खुड्डाकपुष्पं वल्लरि: कुक्षिसंज्ञका ।।

जालकं पञ्चकर्णञ्च स्वैरपञ्चदलाग्रकम् ।
दलानां सन्निवेशञ्च कर्णान्तरे तु जालके ।।
पञ्चैव केशरास्तत्र दलानां पुरोगं भवेत् ।
मुक्तेन वराटेनैव स्थालकस्थं फलं भवेत् ।।
कुड्याख्यबीजाधारश्च वर्तुलो मञ्जुको निभः ।
कर्परं बीजकीखोसं वर्तकं नैकसम्मतम् ।।
वृन्तान्तिकबीजञ्चैकं बीजञ्च वर्तके तथा।
एतानि लक्षणानि हि भवन्ति कुहगणीये ।।७१।।

ओड्रपुष्पगणीयम्

ओड्रपुष्पंप्रवक्ष्यामि भाषितं वाटयमण्डलम् ।
कुक्षिपुष्पं भवेत्तत्र मिश्रमञ्जरिका क्वचित् ।।
जालकमुपपक्षेण युक्तश्च पञ्चकर्णितम् ।
संस्पृष्टपञ्चदलञ्च युक्तं वराटसङ्गमे ।।
बहवः केशास्तत्र वराटोत्सङ्गसंज्ञकाः ।
संवृतञ्च भवेत्तैश्च बीजाधारवराटकम् ।।
संवृतबीजाधारश्च तस्मात् संज्ञायते द्विज ।
वराटं पञ्चशीर्षञ्च केशरैः परिवारितम् ।।
विदरो बीजाधारश्च पञ्चभिर्वर्तकैः सह ।
एतानि लक्षणानि हि भवन्ति पुष्पमण्डले ।।७२।।

मल्लिकागणीयम्

मल्लिका कुण्डमण्डला संकुला पत्रमञ्जरिः ।
अरापक्षपङ्क्तिवृन्ता समाङ्गा सक्षीरा स्मृता ।।
जालकं युक्तकर्णञ्च पञ्च भवति तत्र वै।
पञ्चयुक्तदलञ्चैवाव्यक्तकेशरपञ्चभिः ।।
वराटद्वययुक्ता च युग्मफला भवच्च स्ग ।
आस्फोटा बहुबीजा तुलपुच्छसमन्विता ।।७३।।

वज्रपुष्पगणीयम्

अथातः संप्रवक्ष्यामि वज्रपुष्पस्य लक्षणम् ।
कुक्षिपुष्पं भवेत्तत्र पुष्पञ्च कुण्डमण्डलम् ।।
नासापुटं समं कुण्डं दलाग्रं विषमं भवेत् ।
चत्वारः केशरास्तत्र विषमतुङ्गशीर्षकाः ।।
युग्मवराटकं युक्तं शीर्षेण तु द्विकर्णितम् ।
द्विपुटबीजाधारश्च पुष्पक्रान्तेन दृश्यते ।।
बीजानि बहुनि तत्र द्विपङ्क्तया समीवद् भवेत ।
फलञ्चैव विदरञ्च भवति पुष्पमण्डले ।।७४।।

भृङ्गपुष्पगणीयम्

अथातो भृङ्गपुष्पञ्च भवनि कुण्डमण्डलम् ।
जालकं स्थिरयुक्तञ्च पञ्चकर्णेन विद्यते ।।
भृङ्गपक्षोदरं निभं दृश्यते कर्णितदलम् ।
चत्वारः केशरास्तत्र पुष्पकुण्डे भवन्ति हि ।।
दलाद्विभागदीर्घञ्च नताग्रस्फुटकेशरम् ।
तथा द्विशीर्षवराटं पुष्पक्रान्तं बीजाशयम् ।।
वर्तकमेकबीजञ्च बन्धनं वृन्तमुखतः ।
कुड्याख्यबीजाधारश्च विषमपुष्पमण्डलम् ।।७५।।

कोटरपुष्पगणीयम्

अथातः कोटरपुष्पं भवति कुण्डमण्डलम् ।
कुक्षिमञ्जरिका चैव पुष्पन्तु सफलं समम् ।।
जालकं पञ्चमुक्तञ्च स्थिरञ्च वृन्तसंहतम् ।
घन्टानिभं दलं युक्तं केशराङ्केन संज्ञितम् ।।
व्यक्ततुङ्गकेशरैश्च पञ्चभिः परिशोभितम् ।
द्विर्शीषेण वराटेन भवति पुष्पमण्डलम् ।।
स्थालकस्थो बीजकोषस्त्रिपुटो द्विपुटो भवेत ।
कोटरं द्विविधं विद्याद् विदरं कुड्यसंज्ञकम् ।।७६।।

भद्रपुष्पगणीयम्

ततश्च भद्रपुष्पञ्च शृणु तद् मिश्रलक्षणम् ।
जालकं पञ्चमुक्तञ्च दलञ्चैव द्विस्तवकम् ।।
प्रथमस्तवकं मुक्तं पञ्चदलेन विद्यते ।
द्वितीयस्तवकं युक्तं तुङ्गशीर्षेण भिद्यत ।।
वहव: केशरास्तत्र कुण्डाङ्केन निगदिता: ।
व्यक्तकेशरकं तच्च खुल्लकं मिश्रमण्डलम् ।।
शीर्षैकं वराटं विद्यात् फलश्चैकंबीजं तथा ।
तत्र वृन्तान्तिकं बीजं कुड्याख्याफलमेव च ।।७७।।
अत्रैव ये गणा उक्तास्तेषां समाहार: शृणु ।
चूतपुष्पं समपुष्पं यच्चैव तुन्दमण्डलम् ।।
साम्यं भवति संख्यायां यतश्च दलकेशरम ।।७८।।
विषमं समिपुष्पञ्च विज्ञेयं तुन्दमण्डलम् ।
न तत्र भवति साम्यं संख्यायां दलकेशरम् ।।७९।।
पुप्लिकं तुन्दमण्डलं तदपि विषमं भवेत् ।
तुन्दञ्च स्वस्तिकपुष्पं विषममण्डलं भवेत् ।।८०।।
न तत्र भवति साम्यं संख्यायां दलकेशरम् ।।८१।।
ततश्च त्रिपुषपुष्पं विषमं कुम्भमण्डलम् ।।८२।।
अवाक्पुष्पं भवेत तत्र समञ्च कुम्भमण्डलम् ।।८३।।
अक्षपुष्पं विषमं च तद् भवेत् कुम्भमण्डलम् ।।८४।।
समञ्च कुहपुष्पञ्च यच्चैव तुङ्गमण्डले ।
साम्यं भवति संख्यायां यतश्च दलकेशरम् ।।८५।।
ओड्रपुष्पं विषमञ्च यच्चैव वाटयमण्डले ।
न तत्र भवति साम्यं संख्यायां दलकेशरम् ।।८६।।
मल्लिकागणीये पुष्पं भवति कुण्डमण्डलम् ।
अव्यक्तकेशरपुष्पं समञ्च पुष्पमण्डलम् ।।८७।।
वज्रपुष्पं विषमञ्च भवति कुण्डमण्डलम् ।
न तत्र भवति साम्यं संख्यायां दलकेशरम ।।८८।।
समञ्च कोटरपुष्पं भवति कुण्डमण्डलम् ।।८९।।

विषमं भृङ्गपुष्पञ्च भवति कुण्डमण्डलम् ।।९०।।
विषमं भद्रपुष्पञ्च यच्चैव मिश्नमण्डलम् ।
प्रकृत्या द्विविधदलं भवति भद्रकगणे ।।९१।।
इति समाहारः प्रोक्तः पुष्पाणां गणसंग्रहे ।।९२।।

भवति चात्र

पुष्पञ्च पुष्पमण्डलं पुष्पाङ्गानि पृथक् पृथक् ।
इत्युक्ता नामरुपाभ्यामस्मिन् पुष्पाङ्गसूत्रीये ।।९३।।

इति पराशरकृते वृक्षायुर्वेदे बीजोत्पत्तिकाण्डे पुष्पाङ्गसूत्रीयो नाम पञ्चमोऽध्यायः ।।९४।।

Bījotpatti Kāṇḍa
Chapter - V
Puṣpāngasūtrīyādhyāya

Parāśara said : I will now narrate the Puṣpāngasūtrīyādhyāya -- the chapter dealing with the flower and its different organs [Verse 1]. Flowers are the forerunner of fruits. Puṣpaṁ, Kusumaṁ, Prasūnaṁ, Sumanaṁ are the synonyms of a flower. [Verse 2].

Puṣpapatraṁ (Bract)

Puṣpapatraṁ (Bract) is the forerunner of a flower. It is also called Prākpuṣpa which means the preflowering leaves. [Verse 3]. It is small in size. [Verse 2].

Kuṭmala (Flower bud)

The flower bud is called Kuṭmala, Kalikā or Koraka. [Verse 4].

Puṣpāngāni (Parts of a flower)

The different parts of a flower are :

'Vallari' (Inflorescence), 'Vṛnta' (Pedicel), 'Jālaka' (Calyx), 'Dala' (Corolla), 'Keśara' (Stamens), 'Parāga' (Pollen), 'Kiñjalka' (Anther), 'Parimala' (Fragrance), 'Makaranda' (Nectar), 'Bījādhāra' (Ovary), 'Varāṭaka' (Style) and 'Sthālaka' (Thalamus). [Verse 5]. Fig. 3.

I will now describe a flower. [Verse 6].

The outermost whorl which covers the floral bud is called Jālaka (Calyx). Enclosed within it is the Puṣpadala, the Corolla. The Corolla encloses the Keśara, (Stamens). Within it lies the Bījādhāra (Ovary); on top of it, the Varāṭaka or the Style is placed, being encircled by the staminal whorl. The style is the seat of seed formation. The Kiñjalka or the Anther is fixed on top of the stamen, containing the pollen grains (Parāga) inside. Anthers may be inserted on the stamens (filaments) in two ways, viz. Prāntasandhitaṁ and Pṛṣṭhasandhitaṁ. In the former the anther is attached to the filament at one end cf. Basifixed. In the latter, the filament is attached at a point on back of the anther cf. Dorsifixed . The fragrance of a flower is oily in nature (Snehaśca - essential oil) and called Parimala. The nectar is called Makaranda. It is stored within the floral cavity. [Verse 6-7].

Vallari (Inflorescence)

The axis or the branch that grows terminally or borne at the leaf axil, and on which the flowers are arranged is called the Vallari (Inflorescence). It is also called Mañjari. There may be eight different types of inflorescence, viz. Palāśa vallari, Pankti vallari, Arakā, Chatrā, Guccha vallari, Sankulā, Otupucchikā and Akṣamañjari. [Verse 8].

Palāsa vallari - It is characterised by the presence of bracts. [Verse 9].

Panktimañjari - The flowers are arranged in rows. [Verse 10].

Arakāmañjari - The pedicels are arranged on the inflorescence axis similarly to that of the spokes of a wheel converging at the centre (cf. Capitate). [Verse 11].

Chatrā mañjari - The pedicels are arranged on the inflorescence axis like the radiating ribs of an umbrella. It is also called Kāravikā (cf. Umbel). [Verse 12].

Gucchā vallari - Several clusters of flowers are borne together on a common axis. [Verse 13].

Sankulā vallari - Flowers are arranged on the axis irregularly without having any definite order (cf. Mixed Inflorescence). [Verse 14].

Otupucchikā - The inflorescence resembles a cat's tail (cf. Catkin) [Verse 15].

Akṣamañjari - The inflorescence is borne at the leaf axil. It will be described in detail in the chapter on stems. [Verse 16].

Again inflorescence .may be divided into two types, viz. Saśākha and Aśākha. It is called Saśākha when the inflorescence axis is branched (cf. Compound Inflorescence) and Aśākha when it is unbranched. [Verse 17].

Vṛnta (Pedicel) - It is referred to in the chapter on leaves. [Verse 18].

Puṣpamaṇḍala sanniveśa (Floral Phyllotaxis)

The mode of arrangement of the calyx, corolla, stamens, etc., on the pedicel is of three different types, viz. Vṛttamaṇḍala Pṛthakamaṇḍala and Miśramaṇḍala. In Vṛttamaṇḍala corolla, stamens are arranged in whorls. In Pṛthakamaṇḍala the floral leaves are arranged in rows. In Miśramaṇḍala these are arranged in a mixed manner (Miśra-mixed). [Verse 19].

In most of the flowers calyx, corolla and stamens are arranged in whorls (Vṛtta), rarely a deviation is observed. [Verse 20].

Jālaka (Calyx)

The Jālaka or the calyx has two other synonyms, viz. Kṣārakaṁ and Vapraṁ. [Verse 21]. It may be of two types, viz. Puṣpāntakajālakam and Sthirajālakaṁ. In Puṣpāntaka type, the calyx withers away and falls off with the formation of fruit.

In case of Sthirajālakam, the calyx does not wither away even after the fruit is formed (cf. Persistent calyx). [Verse 22].

Sometimes the persistent calyx remains firmly attached (Samhata) to the pedicel. Fruits or flowers carrying such a calyx are called Vṛntāka (e.g. Brinjal, *Solanum melongena*). Often the calyx is completely fused with the thalamus and is no longer distinguishable. Such type of calyx is found in Piṇḍapuṣpa. [Verse 23].

Calyx is again of two types, viz. Yuktajālakam and Muktajālakaṁ. In Yuktajālakam, the lobes are united (cf. Gamosepalous). When they are free, the calyx is called Muktajālakaṁ (cf. Polysepalous). [Verse 24].

Calyx often shows resemblance to the corolla in shape, and also shows variation. Sepals or lobes of a calyx in some flowers are of uniform size; while they may be dissimilar in other cases. Often the calyx is skin like in appearance, which serves to protect the flower. Such a calyx is termed 'Ajinajālaka'. The calyx in some flowers assumes the form of a widely opened mouth ('Jṛmbhitaṁ'). It is called 'Jṛmbha Jālaka'. Sometimes the sepals are completely united with one another (cf. Entire). [Verse 25].

Upajālaka (Epicalyx)

In some flowers the pedicels bear small leaves below the calyx. It is called Upajālaka (cf. Epicalyx). [Verse 26].

Mocikaṁ (Spathe)

The plants of the grass family (Tṛṇa), Zingiberaceae (Valvaja) and Araceae (Kaṇḍula) bear an enveloping leaf ('Paṭṭikasamkāśaṁ') which protects the spadix ('Akṣamañjari'). It is called Mocikaṁ (Cf. Spathe bract). [Verse 27].

Puṣpadalaṁ (Corolla)

As the calyx serves to protect the young buds similarly also the corolla gives protection to the stamens and the carpels. [Verse 28]. The corolla may be of four types, viz. Muktadala, Yuktadala, keśarānkadala and Swairadala. [Verse 29].

If the petals are free it is called Muktadala (cf. Polypetalous). When the petals are united it is called Yuktadala (cf. Gamopetalous) [Verse 30].

In Keśarānkadala, the stamens adhere to the corolla (Keśara-stamens, Anka-lap, cf. Epipetalous). In Swairadala stamens are not adherent to the petals and are located in between two sepals as found in Tungapuṣpa. [Verse 31]. In some flowers the petals or the corolla lobes are equal in size. They are called Samadala. In some other flowers the petals or corolla lobes are unequal. They are called Viṣamadala. [Verse 32].

Ākṛtito Puṣpadalaṁ (Forms of Corolla)

The petals in a polypetalous corolla mostly resemble a leaf in shape ('Parṇa samkāśaṁ). In some flowers the petals are modified into hairy structures and are called Lomaśa Puṣpa or Śukapuṣpa. [Verse 33].

In some Gamopetalous flowers, the corolla may be shaped like a bell (cf. Campanulate). In some cases they are urnshaped (cf. Urceolate). Sometimes they are incised on top, (cf. Tubular). Sometimes it resembles a widely opened mouth of a lion, cf. bilabiate ('Jṛmbhitaṁ'). Often it is shaped like the mouth of a pigeon ('Kapotavaktra'). Again sometimes it resembles a nasal cavity ('Nāsāpuṭa Saṁkāśaṁ'). The tubular portion of the gamopetalous corolla is called kuṇḍa (cf. Corolla tube). [Verse 34].

Keśara (Stamens)

The organ which bears anthers is called Keśara (Stamens). Stamens may be of three types, viz. Dalotsanga kesara, Sthālakī and Varāṭotsanga. When the stamens adhere to the petals they are called Dalotsanga (Dala - Petals, Utsanga - adhered, cf. Epipetalous). When the stamens grow directly from the thalamus and are free they are called Sthālakī, (cf. Polyandrous). When the stamens are united together forming a tube round the pistil, it is called Varāṭotsanga (Varāṭa - Style, Utsanga - adhred, cf. Monadelphous).

The stamens may be again divided into two types, viz. Samakeśara and Viṣama keśara. If the stamens are of equal length they are called Samakeśara. If they are of different length they are called Viṣama Keśara. [Verse 35].

Ākṛtito Keśara (The shape of the stamens)

The stamens may be of five types according to the shape, of the anthers, viz. Tungakeśara, Mañjukeśara, Pannagī, Vājīkrānta and Urmikeśara.

In Tungakeśara, the anther is placed at the top of the filament. In Mañjukeśara, the anther is attached to the filament in a beautiful fashion (Mañjulena means beautifully).

In Pannagīkeśara, the anther resembles the hood of a snake. In Vājīkrāntakeśara, the anther is shaped like a horse's hoof. In Urmikeśara, the stamens are united along their edges enclosing the style and assume a wavy appearance - (Urmi' means wave), cf. Staminal corona. [Verse 36].

Bijādhāra (Ovary)

Seeds are formed within the Bijādhāra (Ovary) which also refers to a fruit. It may be of two types, viz. Vidara and Kuḍya. It is called Vidara when the ovary is dehiscent (Sandhita), and Kuḍya when it is indehiscent (Asandhita). [Verse 37].

Bījādhārasya Vartakaṁ (The Locule of the ovary)

An ovary may be unilocular (Ekavartakaṁ) or multilocular (Bahuvartakaṁ). The ovary is divided into locules (Vartaka) by the septa (Puplika) and by the placenta (Puṣa).

Again an ovary may be divided into three types according to the nature of flowers, viz. Puṣpakrānta, Samvṛta and Puṣpaśīrṣaka. In Puṣpakrānta type, the ovary is placed within the floral cavity (cf. Hypogynous).

In Samvṛta type the ovary is enclosed within the tubular portion formed by the union of corolla and the stamens which resembles a hand pounding implement (cf. Perigynous). In Puṣpaśīrṣaka type, the ovary is placed below, and above it the other whorls are inserted (cf. Epigynous). [Verse 38].

Varāṭaka (Style)

The portion on top of the ovary which remains surrounded by the stamens and within which seeds are produced is called the Varāṭaka (Style).

It may have one, two or many stigmas depending on the flower. The stigmas may be separate or united. The style may be long tubular; short or globular (Vartula). The region which connects the ovary and the style is called the Varāṭasangaṁ. [Verse 39].

When pollen grains along with nectar ('Makarandena saha') come in contact with the style, then in due course, seeds are formed. [Verse 40].

Sthālaka (Thalamus)

The axis on which the floral members are borne is called the Sthālaka or the thalamus. Sometimes it remains firmly attached to the pedicel. In some flowers it is visible above the whorl of sepals.

The Sthālaka or the thalamus may be of three types according to the variation of forms viz. Kuṇḍa, Falgu and Piṇḍa.

In Kuṇḍa type, the thalamus is flattened like a disc (śarāba). In Falgu type, it is round and hollow, ('Falgu' means hollow). In Piṇḍa type, it forms a solid round structure ('Piṇḍa' means solid). [Verse 41].

Puṣpa Gaṇavibhāga (Classification of flowers)

Parāśara said : I will now describe the different types of flowers, which are commonly seen, according to their location (on the stem), nature, forms, names and colour. [Verse 42].

According to their location on the stem, flowers may be grouped into three types, viz. Kukṣi puṣpa, Kāṇḍa puṣpa and Vallari puṣpa. [Verse 43].

Kukṣi puṣpa - The flowers that develop singly or in clusters at the leaf axil are termed Kukṣi puṣpa. They are also called Ekapuṣpa Vallari (cf. Solitary). [Verse 44].

Kāṇḍa Puṣpa - The flowers which are borne singly or in clusters on the stem are said to be kāṇḍa puṣpa. [Verse 45].

Vallari Puṣpa - The flowers borne on the inflorescences at the stem terminal are called Vallari puṣpa . [Verse 46].

In some cases the inflorescence bears a large number of sessile flowers arranged on a flat disc like thalamus (Kuṇḍasthālaka) enclosed within the same calyx (cf. Involucre). The inflorescence appears like a brush-head.

The ovary is epigynous (Puṣpaśīrṣāka). The flowers are said to be Kurca Puṣpa (e.g. *Tagetes erecta*). [Verse 47].

Saphala and Niṣphala puṣpa (Fertile and sterile flowers)

Flowers by nature may be fertile or sterile. A flower is called Saphala (fertile) when the ovary is present. When it is absent the flower is called Niṣphala (sterile). As a consequence, the plants also may be fertile or sterile. Rarely both fertile and sterile flowers are borne on the same plant or on the same inflorescence. [Verse 48].

Flowers may have a single ovary or two or many ovaries. [Verse 49].

Pratyak puṣpa

In some types, flower buds in the inflorescence face upwards, but when they blossom they change their direction (e.g. *Achyranthes aspera*) [Verse 50].

Vyakta and Avyakta Puṣpa (Open and Closed flowers)

Flowers are called Vyakta or open when the stamens protrude out (cf. Exserted) with the opening of the petals. Flowers are called Avyakta or closed when the stamens remain enclosed within the corolla tube and are not visible. (cf. Inserted). [Verse 51].

Ravikānta and Candrakānta Puṣpa

Flowers again may be divided into two types according to the dominating influence of the sun or the moon upon them. Flowers which bloom at night when the moon has its influence, are called Candrakānta puṣpa (Candra - the moon). Flowers which bloom during the day when the sun dominates are called Ravikānta puṣpa (Ravi - the sun). [Verse 52].

Ṛtu Puṣpa and Sadā Puṣpa

The flowers which bloom at a particualr time of the year are called Ṛtu Puṣpa (seasonal flowers). Flowerș which bloom throughout the year are called Sadā Puṣpa. [Verse 53].

Flowers may again be classified as Vyakta puṣpa and Avyakta puṣpa. It is called Vyakta puṣpa when the flowers are fully opened exposing the petals, stamens and the styles (cf. Chasmogamy). It is called Avyakta puṣpa when the petals, stamens and the styles remain enclosed within a thalamus which is round and hollow (Falgu sthālaka) and is thus invisible. [Verse 54].

Muktadala Yuktadala Puṣpa (cf. Polypetalous and Gamopetalous Corolla)

Flowers are called Muktadala (cf. Polypetalous) when the petals are free from one another; and Yuktadala (cf. Gamopetalous) when they are united together. [Verse 55].

Sanspṛṣṭa Dala and Sansphuṭa Dala (cf. Twisted and Valvate Aestivation)

Again flowers may be of two types, viz. Sanspṛṣṭa dala and Sansphuṭa dala. A flower is termed Sanspṛṣṭa when the petals are arranged embracing (Śliṣṭa) one another, (cf. Twisted aestivation). It is called Sansphuṭa if the petals are separate from one another without any overlapping, (cf. Valvate aestivation). [Verse 56].

Samānga and Hīnānga Puṣpa (cf. Complete and Incomplete flower)

A flower is called Samānga when the calyx, corolla, stamens and carpels, all four sets of members are present (cf. Complete flower). A flower is called Hīnānga when any of the floral whorls is missing (cf. Incomplete flower).

The Hīnānga puṣpa may be of two types. In the former type, carpels are absent and it is called Mañju puṣpa or Keśara puṣpa. In the latter, petals and stamens are absent and the flower is called Raṇḍa puṣpa or Śrīpuṣpa (cf. female flower). The Mañju puṣpa is, naturally, sterile, as the ovary is absent. Rarely both male and female flowers are developed on the same plant or on the same inflorescence. Quite often, in Śrīpuṣpa or Raṇḍa puṣpa (female flower) the ovary is trilocular with a style bearing three stigmas. If male and female flowers are borne on the same inflorescence it is called Mañjuśrī puṣpa (cf. Monoecious). [Verse 57].

Bījādhāra Vyanjanena Puṣpamaṇḍala (Insertion of floral leaves on the thalamus)

According to the position of ovary (on the thalamus) the flowers may be divided into four classes, viz. Tundamaṇḍala, Kumbhamaṇḍala, Tungamaṇḍala and Vāṭyamaṇḍala.

When the petals and stamens are inserted below the ovary, the flower is called Tundamaṇḍala (cf. Hypogynous). The ovary is called Puṣpakrānta (cf. Superior). If the petals and stamens are placed above the ovary, the flower is termed Kumbhamaṇḍala (cf. Epigynous). The ovary is called Puṣpaśīrṣaka (cf. Inferior). Such a condition may be found in both polypetalous and gamopetalous corolla. When the sepals, petals and the epipetalous (Dalotsanga) stamens are inserted on a circular disc-like thalamus surrounding the style at the centre, and the ovary being fused with the thalamus, the flower is called Tungamaṇḍala (cf. Perigynous). When the petals, stamens are united forming a column round the style it is called Vāṭyamaṇḍala. [Verse 58] Fig. 4a-d

Sama-Viṣama Gaṇīya Puṣpa (cf. Isomerous and Heteromerous flowers)

Flowers may be classified into Samagaṇīya or Viṣamagaṇīya. When in a flower there are equal number of petals and stamens, it is called Samagaṇīya (cf. Isomerous). When the number is unequal it is called Viṣamagaṇīya (cf. Heteromerous). [Verse 59].

A flower with five petals or with corolla divided into five lobes is a common occurrence; a flower having several whorls of petals arranged successively is called Stabaka Puṣpa. They are arranged in an inner and outer whorls. [Verse 60-61].

Varṇataśca Puṣpāṇi (Colour of the flowers)

Flowers are found in various shades of colours - white, red, golden, dark brown and blue. Plants are characterised by the colour of their blossoms. Flowers as found in the vast population of trees, shrubs and creepers show a great variety in forms, shape, names, sex, etc. The floral characteristics of some of the families will now be described here. Those which have not been mentioned here will be described later in the respective chapters. [Verse 62-63].

CŪTAGAṆĪYA PUṢPA (cf. Anacardiaceae)

Petals and stamens are free from one another; flowers are small and hypogynous; inflorescence is branched (Saśākha), coming out in spring (Madhumañjari); petals and sepals are five each; stamens five, bearing short reddish anthers; styles free ('Uchrita'); fruit is hypogynous (Puṣpakrānta), indehiscent (Kuḍya), containing a single seed; seed coat is hard. Its attachment is Vṛntāntika (oriented towards pedicel). [Verse 64].

SAMĪGAṆĪYA PUṢPA (cf. Leguminosae)

In this family flowers are hypogynous (Tundamaṇḍala), having petals of different shapes (Viṣama Vidala); petals are five and free; sepals united, lobed; stamens ten (a characteristic of Samīpuṣpa). Fruits are legumes. Seeds are borne in lateral rows ('Pārśwabīja'). According to the variation in forms, flowers are called Vakrapuṣpa, Vikarṇikapuṣpa and Śūkapuṣpa. (cf. Papilionaceae, Caesalpiniaceae and Mimosaceae respectively). [Verse 65].

PUPLIKGANĪYA PUṢPA (cf. Rutaceae)

Petals and sepals are separate; flower hypogynous ('Tundamaṇḍala'); heteromerous (Viṣama); inflorescences come out in spring; sepals five; petals five; stamens many, attached to the base of the anther ('Prāntagranthikena', cf. Basifixed), styles free with many stigmas; fruits indehiscent; edible portion of the fruit (Phalaśasya) consists of fibrous outgrowths resembling the stamens (Keśaranibham); fruit hypogynous (Puṣpakrānta), divided into many chambers, each having several seeds. [Verse 66].

SWASTIKGAṆĪYA PUṢPA (cf. Cruciferae)

Parāśara said : I will now describe the distinguishing characters of the Swastikgaṇīya Puṣpa. Flowers are hypogynous, borne in rows on the inflorescence: form of the calyx resembles the symbol of a 'Swastikā'. Flowers are polysepalous; petals four, polypetalous; stamens six free of which two are shorter (cf. Tetradynamous); styles two, united; fruits having two chambers; fruits are dehiscent, similar to leguminous fruits. [Verse 67].

N.B. Ovary actually one chambered, but appears two-chambered because of the development of a false partition wall. (cf. Replum).

TRIPUṢAGAṆĪYA PUṢPA (cf. Cucurbitaceae)

Parāśara said : I will now describe the distinguishing characters of the Tripuṣagaṇīya puṣpa.

Flowers are epigynous; some flowers are sterile (Niṣphala), usually borne in the axils of leaves, rarely a compound inflorescence; sepals and petals both united, with five lobes each; stamens three, styles with three stigmas; ovary trilocular, divided by three placental tissue (Puṣa); with many seeds in each locule; seeds are flat. [Verse 68].

AVĀKPUṢPAGAṆĪYA (cf. Umbelliferae)

Inflorescence is an umbel (Chatramañjari); The pedicels are arranged like the radiating ribs of an umbrella called 'Kārava' and hence the inflorescence is also termed 'Kāravikā'. Petals five, minute, polypetalous; stamens five, very fine; flower epigynous (Kumbhamaṇḍala); styles two, ovary bilocular; ovary is called 'Bījaka'. [Verse 69].

AKṢAPUṢPAGAṆĪYA (cf. Combretaceae)

Parāśara said : I will now describe the Akṣapuṣpa. In this family the inflorescence is branched; flowers small, epigynous; sepals are united with five lobes; petals five; stamens ten; style one; fruits are indehiscent (Kuḍya). Seeds are brittle. Seeds are also termed 'Asrak' (with ridges) and 'Kīkhosaṁ'. Fruit contains a single seed. The fruit is 'Vṛntāntika'. (oriented towards the pedicel). [Verse 70].

KUHAPUṢPAGAṆĪYA (cf. Rhamnaceae)

Parāśara said : I will now describe the Kuhapuṣpa. Flowers are perigynous, complete, small; inflorescences are axillary (Kukṣi vallari); sepals are united with five lobes; petals five; they are arranged in between two sepals; stamens five, placed in front of the petals; style free; ovary indehiscent and round; seed coat is hard. The fruit contains a single seed which is Vṛntāntika (oriented towards the pedicel). [Verse 71].

OḌRAPUṢPAGAṆĪYA (cf. Malvaceae)

Parāśara said : I will now describe the Oḍrapuṣpa. Flower is termed Vāṭyamaṇḍalīya. Stamens are united, forming a column round the style. Flowers are borne in the axils of leaves, sometimes they are of mixed inflorescence (cf. Compound Cyme); calyx five, lobed; epicalyx (Upajālaka) present; petals five, united below at the juncture of the ovary and the style; stamens many, united forming a tube enclosing the style and the ovary. As the ovary and the style are surrounded by the staminal tube, the flower is called Saṁvṛta. The style with five stigmas; ovary dehiscent, five loculaṛ [Verse 72].

MALLIKĀPUṢPAGAṆĪYA (cf. Apocynaceae)

Flowers with disc like thalamus (Kuṇḍamaṇḍalīya); leaves are irregularly arranged on the inflorescence axis; leaves are arranged on the stem like the spokes of a wheel; flowers are complete (Samānga); plants with latex (Sakṣīrā); sepals united with five lobes; petals united, five; stamens five, inserted (Avyakta Keśara); styles two, united; fruits borne in pairs (Yugma Phala cf. Follicles); seeds having hairy out growth like cotton fibres (Tulapuccha). [Verse 73].

VAJRAPUṢPAGAṆĪYA (cf. Pedaliaceae)

Parāśara said: I will now describe the distinguishing characters of the Vajrapuṣpagaṇa. Flowers are developed in the axils of leaves; flowers form a deep cavity inside (Kuṇḍamaṇḍalīya); cavity is as deep as the nasal passage (Nāsāpuṭāsama). Petals are of varying size; stamens four of unequal length; styles two united, with bifid stigmas; ovary superior (Puṣpakrānta), bilocular; seeds are arranged in two rows like that of the leguminous fruits; fruit dehiscent (Vidara). [Verse 74].

BHṚNGAPUṢPAGAṆĪYA (cf. Verbenaceae)

Parāśara said: I will now describe the distinguishing characters of the Bhṛngapuṣpa. Flower with a deep cavity inside (Kuṇḍamaṇḍalīya) calyx persistent (Sthirajālaka); sepals united with five lobes; petals united; shaped like the abdomen with wings of a wasp (Bhṛnga); stamens four, inside the tubular cavity; twice the length of the petal; filaments bent at the apex; stamens free from the petals; long; styles bifid; ovary superior (Puṣpakrānta), unilocular with a single seed, Vṛntāntika (oriented towards the pedicel); flower heteromerous (Viṣama); ovary indehiscent (Kuḍyam). [Verse 75].

KOṬARAPUṢPAGAṆĪYA (cf. Convolvulaceae)

Flower having a tubular structure (Kuṇḍamaṇḍalīya); inflorescence borne at the axils of leaves; flowers are isomerous, complete; sepals five, free, attached to the pedicel, persistent (Sthirajālaka); corolla five, united, bell-shaped, with epipetalous stamens (characteristic of the family); stamens five, anthers placed on top of the filaments (Tungakeśara); styles bifid, ovary bilocular or trilocular; fruits may be dehiscent or indehiscent (Vidaram, or Kuḍyam). [Verse 76].

BHADRAPUṢPAGAṆĪYA (cf. Meliaceae)

Parāśara said : I will now describe the distinguishing characters of the Bhadrapuṣpa. Flowers are Miśramaṇḍala. Sepals five, free (Polysepalous - 'jālakampañca muktanca'). Petals are in two whorls, of which the outer whorl consists of five free petals. The inner whorl has gamopetalous corolla. Stamens are many, located inside the corolla tube ('kuṇḍānkena nigaditah'). Style one; the fruit has got only one seed oriented towards the pedicel; the fruit is indehiscent. [Verse 77].

Parāśara said : I will now give the summary of the each individual flower type described:

CŪTAPUṢPAGAṆĪYA (cf. Anacardiaceae)

Flower complete, hypogynous and isomerous. [Verse 78].

SAMĪPUṢPAGAṆĪYA (cf. Leguminosae)

Flower heteromerous, hypogynous. [Verse 79].

PUPLIKAGAṆĪYA PUṢPA (cf. Rutaceae)

Flower, heteromerous, hypogynous. [Verse 80].

SWASTIKAGAṆĪYA PUṢPA (cf. Cruciferae)

Flower heteromerous, hypogynous. [Verse 81].

TRIPUṢAGAṆĪYA PUṢPA (cf. Cucurbitaceae)

Flower epigynous, isomerous. [Verse 82].

AVĀKPUṢPAGAṆĪYA (cf. Umbelliferae)

Flower epigynous, isomerous. [Verse 83].

AKṢAPUṢPAGAṆĪYA (cf. Combretaceae)

Flower epigynous, heteromerous. [Verse 84].

KUHAPUṢPAGAṆĪYA (cf. Rhamnaceae)

Flower perigynous, isomerous. [Verse 85].

OḌRAPUṢPAGAṆĪYA (cf. Malvaceae)

Flower with stamens being united, forming a tube enclosing the style (Vāṭyamaṇḍalīya); heteromerous. [Verse 86].

MALLIKĀGAṆĪYA PUṢPA (cf. Apocynaceae)

Flower forming a tubular cavity, isomerous stamens inserted. [Vere 87].

VAJRAPUṢPAGAṆĪYA (cf. Pedaliaceae)

Flower forming a tubular cavity; heteromerous. [Verse 88].

KOṬARAPUṢPAGAṆĪYA (cf. Convolvulaceae)

Flower forming a tubular cavity, isomerous. [Verse 89].

BHṚNGAPUṢPAGAṆĪYA (cf. Verbenaceae)

Flower forming a tubular cavity, heteromerous. [Verse 90].

BHADRAPUṢPAGAṆĪYA (cf. Meliaceae)

Flower 'Miśramaṇḍalīya', heteromerous, petals are arranged in two whorls. [Verse 91].

Thus ends the summary of the different families of flowers. [Verse 92].

Summary

Parāśara said : Different parts of the flowers viz, Puṣpavṛnta (Pedicel), Sthālaka (Thalamus), Jālaka (Calyx), Dala (Corolla), Kiñjalka (Stamen), Varāṭaka (Style), Bījādhāra (Ovary), and the arrangement of the floral members as well as the names, shapes, forms of flowers have been described in the Puṣpāngasūtrīyādhyāya, the fifth chapter of the Bījotpatti Kāṇḍa of the Vṛkṣāyurveda by Parāśara. [Verse 93-94].

बीजोत्पत्तिकाण्डः
फलाङ्गसूत्रीयाध्यायः

अथातः फलाङ्गसूत्रीयमध्यायं व्याख्यास्याम इत्याह पराशरः ।।१।।

खलु फलञ्च बीजाधारो बीजपुटक इत्यनर्थान्तरम् ।।२।।

फलानां गणविभागं पुष्पतः प्रकृतितश्चाकृतितो नामतो रसतश्च प्रत्यक्षेण यदुपलभ्यते तदनुव्याख्यास्यामः ।।३।।

इह खलु वृक्षाणां बीजाधारः फलञ्च पुष्पतस्त्रिविधं विद्यात् । पुष्पक्रान्तं पुष्पशीर्षकञ्च संवृतञ्चेति ।।४।।

तत्र पुष्पक्रान्तबीजाधारो जालकस्योपरि पुष्पकुक्षौ सन्निविष्टं भवति। यथाम्र बिल्वबदरसमीफलादीनि । पुष्पक्रान्तफलन्तु किञ्चिदेकबीजं किञ्चिद् बहुबीजं वा भवति ।।५।।

पुष्पशीर्षकस्य तु बीजाधारशीर्ष पुष्पसन्निविष्टं भवति। यथा दाडिमपालेवत–तुम्वित्रपुषजालिनीफलादीनि पुष्पशीर्षक फलानि स्युः। पुष्पशीर्षकफलानि बहुवर्तकानि प्रायशो बहुवीजानि भवन्ति ।।६।। संवृतबीजाधारस्तु पुष्पमण्डले यच्च दलकेशरसंधाताङ्गेन संवृतं भवति। संवृतफलं बहुवर्त्तकं बहुबीजञ्च भवति ।।७।।

खलु वृन्तं जालकं वल्कं शलाटु वर्तकं बीजपुप्लिकं बीजपुषं बीजञ्चेति फलाङ्गानि भवन्ति ।।८।।

तत्र वृन्तजालकौ पुष्पाङ्गसूत्रीये वक्ष्यते ।।९।।

खलु फलस्य बहिरावरणं वल्कं वल्कलञ्चेति। फलवल्कञ्च मृदुलमालुरांशुकशूकाचितकण्टककीलकार्वुदाचितभेदेन भिद्यते ।।१०।।

भवतश्चात्र।

नखच्छेद्यं मृदुलञ्च मालुरं कर्परं स्मृतम् ।
अंशुकं तन्तुलं विद्याच्छुकव्यासं शूकाचितम् ।।
कण्टकैः कण्टकं ज्ञेयं कीलकैः कीलकं स्मृतम् ।
अर्वुदैरर्वुदं विद्यादित्युक्तं फलवल्कलम् ।।११।।

धात्रीलकुचद्राक्षाफलादीनि मृदुलवल्कलानि। विल्वकपित्थफलादीनि मालुरवल्कलानि ।तृणफलादीनि प्रायशश्चांशुकवल्कलानि भवन्ति ।।१२।।

द्विविधं हि पुनः फलवल्कलञ्च सन्धितासन्धितभेदेन ।सन्धितं पुनरेव द्विविधं तद् यथाऽऽदीर्घसन्धितं मण्डलसन्धितञ्च ।आदीर्घसन्धितं यच्चैव ऋजुक्रमेण सन्धितं वृन्ताद् यावच्छीर्षम्। यथास्फोटकसमिधात्र्याम्लवेतसफलादीनि आदीर्घसन्धितवल्कलानि भवन्ति ।।१३।। यच्चैव फलशीर्षे वृत्तमण्डलेन सन्धिना सन्धितं येषां फलपरिपक्के शुष्के च कृतच्छिद्रं भवति बीजापसरणाय तच्चैव मण्डलसन्धितवल्कलम् । यथा जालिनीफलादीनि कृतवेधनानि । सन्धितवल्कलञ्च विदरफलेषु दृश्यते ।।१४।। असन्धितं वल्कलञ्च यदखण्डेन दृश्यते ।यथा जम्बाम्रराजादनकुड्यफलवल्कलानि ।।१५।।

खण्डशलाटुवल्कलञ्च यद् शलादुना सह विभक्तञ्च यथा धात्र्याम्लवेतसफलौ तौ शस्येन सह आदीर्घसन्धितौ ।।१६।। खलु फलवल्कलाभ्यन्तरे फलशस्यं शलाटु यच्चैव शुष्के वानसंज्ञकमाचक्षते ।।१७।।

वर्तकन्तु बीजानामाशग्रः पुटकश्च ।।१८।।

ततो बीजपुप्लिकं बीजपुरञ्च विज्ञायते येन फलकुक्षि विभक्तं भवति । यच्च फलभ्यन्तरे निवहबीजानामावरणञ्च । तच्चैव द्विविधं मृदुकठोरभेदाच्च । बीजपुषन्तु यच्चैव केषाञ्चित् फलानां निवहबीजानां संवृतिकञ्च। एवञ्च यत्र हि बीजानि संसक्तानि भवन्ति ।।१९।। बीजञ्च बीजाङ्गसूत्रीये वक्ष्यते।।२०।।

खलु फलकुक्षौ पञ्चविधं बीजबन्धनं भवति । यथा वृन्तान्तिकञ्च पुप्लिकान्तिकञ्च पुषान्तिकञ्च स्थालकान्तिकञ्च पङ्क्तिकञ्चेति । तत्रैकवर्तके वृत्तान्तिकबन्धनन्तु वृन्तमभितो बीजमुखं भवति । पुप्लिकान्तिकं नाम बन्धनं यत् पुप्लिकेन संवृतं भवति । पुषान्तिकं नाम बन्धनं यत् पुषेण संवृतं संसक्तञ्च भवति । स्थालकान्तिकं नाम बन्धनं यत् स्थालके दृश्यते । पङ्क्तिवन्धनन्तु पङ्क्त्या पार्श्वेन यद् भवति ।।२१।।

प्रकृत्या परिपक्वे शुष्के वा येभ्यः सन्धितवल्कलफलेभ्यः सश्ब्देन दूरे बीजानि निपतन्ति तानि फलानि नामतश्चाभ्रबीजफलानि भवन्ति । यथा कुरुण्टादयश्चेति ।।२२।।

प्रकतितः फलानां गणविभागश्चतुर्विधो भवति । तद् यथा तुन्दिकं बीजकं कुर्चमास्फोटकञ्चेति ।।२३।।

तत्र तुन्दिकगणीयानां फलानां स्वातन्त्रबीजत्वेन तेषां फलशस्यमन्तरेण बीजं प्ररोहति। यथाम्रजम्बुधात्रीबदरबिम्बिकापॉसादीनि तुन्दिकगणीयाति भवन्ति ।।२४।।

बीजकन्तु तथा विधौ न च प्ररोहति फलबीजत्वयोरभेदात् यथा शाककककुभसिन्धुवारकारविकातृणधान्यादीनि ।।२५।।

कुर्चगणीयानि तु बहुफलान्वितेन कुर्चाकारेणैकस्थालकस्योपरि सन्निविष्टानि स्युः । कुर्चफलानि पुष्पशीर्षकाणि चैव प्रायशः परिपक्केषु स्थालकात् परिच्युतं भवन्ति ।।२६।।

कुर्चफलगणीयन्तु पुनर्द्विविधम् पुच्छिकमपुच्छिकञ्च । तत्र पुच्छिकं यच्चैव बीजपक्के स्थालकस्योपरि छत्राकारं भवत्येव वायुना सञ्चरते रोमपुच्छयुक्तत्वात्। यथा सोमराजादयश्च ।।२७।। अपुच्छिकन्तु पुच्छमन्तरेण (Lacuna exists) यथा केशराजादयश्च । एवंविधानि यानि चान्यानि तान्यपुच्छिकानि ।।२८।।

आस्फोटकगणीयस्य पुष्पमण्डलं द्विफलीयं द्विवराटकञ्च भवति । क्वचिदुर्मिकेशरपुष्पेण विद्यात् एवंविधानि यानि क्षीरवृक्षजानि

तान्यास्फोटकगणीयानि भवन्ति । एतेषां बीजञ्च तुलपुच्छिकं यच्च वायुना सञ्चरते। एवमेकेन पार्श्वेन फलवल्कलमादीर्घेसन्धितं च भवति यथा अर्कादयश्च एवंविधानि यानि तान्यास्फोटकगणीयानि ।।२९।।

अत ऊर्ध्वं फलानां गणविभागमाकृतितश्चोपदेक्ष्यते। तद् यथा आवर्तुलञ्च वर्तुलञ्च दीर्घञ्च पृथुकञ्चेति फलानामाकृतिमादिशेत्। तत्रानुक्तेषु मिश्नलक्षणेषु मिश्नलक्षणमेवोपलब्ध्वा मिश्नाकृति व्यवस्येत् ।।३०।।

भवन्ति चात्र।

नामतश्च प्रवक्ष्यामि फलानां गणसंग्रहम् ।
फल्गुफलं कुम्भफलं त्रिपुषं जालिनी तथा ।।३१।।
भस्त्रा शृङ्गी तुलिनी च त्रिपुटा वृन्ताकी तथा
गुच्छं गुच्छदन्तिका च केशरकफलं स्मृतम् ।।३२।।
कोषंशङ्कुसिम्बिफलं बीजकफलमेव च।
एतैश्च नामगणैश्च फलानां गणसंग्रहः ।।३३।।
ततो नामगणै र्यानि चोक्तानि फलसंग्रहः ।
तानि पुनः प्रवक्ष्यन्ते प्रकृत्यनुरुपैश्च वै ।।३४।।
फल्गुफलं भवेत्तच्च यत्र स्यात् फल्गुस्थालकम् ।
तच्च वनस्पतिकाण्डे व्यासत उपदेक्ष्यते ।।३५।।
कुम्भवत् कुम्भफलञ्च त्रिपुषैस्त्रिपुषं स्मृतम् ।
वानस्तु जालवद् यस्या जालिनी सा च कीर्तिता ।।३६।।
भस्त्रैव पुटकाकारमजिनमिव जालकम् ।
तेनैव वेष्टिता या तु भस्त्राफला सा कीर्तिता ।।३७।।
शृङ्गेन सदृशं शृङ्गी तुलिनी तुलगर्भिणी ।
त्रिभिः पुटैस्तु विभक्ता त्रिपुटा चैव कीर्तिता ।।३८।।
अभेदं वृन्तजालकं सा चैव वृन्ताकी स्मृता ।।३९।।
बहुभिः फलयुक्तैश्च वृन्ते यद् गुच्छबन्धनम् ।
तदेव गुच्छफलञ्च कथितं मुनिपुङ्गवैः ।।४०।।

दन्तपङ्क्तिसमा या तु विज्ञेया गुच्छदन्तिका ।।४१।।
शलादु केशराकारं विभक्तं बहुपुप्लिकैः ।
फलं केशरकं ज्ञेयं क्वचित्तत्र विशेष्यते ।।४२।।
वल्कलाभ्यन्तरे यच्च स्थालके पिण्डसंज्ञके ।
तच्चैव कोषफलञ्च बहुकोषबीजान्वितैः ।।४३।।
शङ्खुवत् शङ्खुफलञ्च पार्श्वबीजैः समिः स्मृतः ।
आदीर्घेण द्विपार्श्वेण सन्धितं सिम्बिवल्कलम् ।।४४।।
अभेदं फलबीजञ्च बीजकं तच्च विज्ञेयम् ।।४५।।
इति संक्षेपतः प्रोक्तः फलानां गणसंग्रहः ।।४६।।
अथ रसतः पुनश्च मधुराम्लतिक्तकटुकषायाश्चेति फलानां रसगणो पञ्च प्रविभागो रसप्राधान्यात् । अपरिसंख्योनां पुनरेतेषां फलानामाश्रयादीनां पाञ्चभौतिकभावानां न्यूनातिरेकान्मिश्रीभावविशेषाच्च कालप्रकर्षाच्च पक्कापक्केषु। फलेषु वैषम्यापद्यते ।।४७।।

इति पराशरकृते वृक्षायुर्वेदे बीजोत्पत्तिकाण्डे फलाङ्गसूत्रीयो नाम षष्ठोऽध्यायः ।।४८।।

Bījotapatti Kāṇḍa
Chapter - VI
Phalāngasūtrīyādhyāya

Parāśara said : I will now narrate the Phalāngasūtrīyādhyāya, the chapter dealing with fruits. [Verse 1].

Synonyms for fruit : Phalam, Bījādhāra, Bījapuṭaka are synonymous. [Verse 2].

I will now give an account of classification of fruits by their names, by the character of the juices they produce as well as by the nature and shapes of their flowers as observed in nature. [Verse 3].

Based on floral characters, fruits may be of three types: Puṣpakrānta, Puṣpaśīrṣak and Saṁvṛta. [Verse 4].

Puṣpakrānta phalaṁ

The ovary is placed within the floral cavity, above the whorl of sepals, e.g. Āmra (*Mangifera indica*); Bilwa (*Aegle marmelos*); Kadara (*Acacia* sp.) and legumes. The fruit may have single seed or have many seeds. [Verse 5].

Puṣpaśīrṣak Phalaṁ

The ovary is placed below the other floral members, e.g. Dādima (*Punica granatum*); Pālevata (*Psidium guajava*); Tumbi (*Lagenaria* sp.); Trapuṣa (*Cucurbita* sp.); Jālinī (*Luffa cylindrica*). Generally Puṣpaśīrṣak fruits are multilocular and may contain many seeds. [Verse 6].

Samvṛta Bījādhāra

The ovary is surrounded by the whorl of stamens and petals united together. The fruit consists of several chambers, each with many seeds. [Verse 7].

Phalāngāni (Parts of a fruit)

The different parts of a fruit are : Vṛnta (Peduncle/ Pedicel). Jālaka (Calyx), Valkaṁ (Fruit wall), Śalāṭu (unripe portion), Vartaka (Locule), Bījapuplika (Septum), Bījapuṣa (Placenta) and Bīja (Seed). [Verse 8].

Of these, Vṛnta (Peduncle/Pedicel) and Jālaka (Calyx) have already been described in the Puṣpangasūtrīyādhyāya (chapter V). [Verse 9].

Phalavalkalaṁ (Fruit wall)

Phalavalkalaṁ is the outermost layer of the fruit (Fruit wall). It may be Mṛdula, Mālura, Angśuka, Śukāvṛta 'Kaṇṭaka' Kīlaka and Arvuda. [Verse 10].

Mṛdula - The fruit wall is soft and could be incised by a nail.

Mālura - The fruit wall is hard and brittle.

Angśuka - The fruit wall is made up fibrous tissue.

Śūkācitam - The fruit wall is covered with hairs.

Kaṇṭaka - The fruit wall is covered with spines (e.g. *Datura metel*).

Kīlaka - The fruit wall is covered with short pointed projections.

Arvuda - The fruit wall is irregularly shaped. [Verse 11].

Dhātrī (*Phyllanthus embilca*), Lakuca (*Artocarpus lakoocha*). Drākṣā (*Vitis vinifera*) fruits are of Mṛdula type; Vilwa (*Aegle marmelos*), Kapittha (*Feronia elephantum*) are of Mālura type. Tṛṇaphala (Fruits of the grass family) are of Angśuka type. [Verse 12].

Sandhita and Asandhita Phalavalkala

(Dehiscent and Indehiscent Eruit Wall)

Phalavalkala or fruit wall may be Sandhita (Dehiscent) or Asandhita (Indehiscent). Again the Sandhita type may be divided into two groups, viz. Ādīrgha Sandhita and Maṇḍala Sandhita. In case of Ādhīrgha Sandhita type, the fruit dehisces longitudinally from pedicel to the apex e.g. Āsphotakaṁ, Samī phala (Legumes), Dhātrī (*Phyllanthus emblica*), Amla vetasa (*Garcinia pedunculata*). [Verse 13].

In case of Maṇḍala Sandhita type, the fruit dehisces transversely or circularly at the top (cf. Circumcissile). When the fruit is mature and dry it splits open at the top releasing the seeds, e.g. Kṛtavedhana, Jālinī (e.g. *Luffa cylindrica*). (Verse 14].

The Asandhita or Indehiscent fruits do not split open at maturity, e.g. Jambu (*Syzygium* sp.) Āmra (*Mangifera indica*) Rājādan (*Alstonia venenata*) [Verse 15].

Khaṇḍa Śalāṭu Valkala

In this particular type of dehiscence, the fruit wall and the unripe portion of the fruit, breaks up longitudinally into several segments, e.g. Dhātrī (*Phyllanthus emblica*), Amlavetasa (*Garcinia pedunculata*). [Verse 16].

The unripe portion inside the fruit is called Śalāṭu. The dry fruits are called Vāna. [Verse 17].

Vartakaṁ (Chambers of the fruit)

Vartakaṁ is the chamber within the fruit enclosing seeds. [Verse 18].

Bījapuplikam and Bījapuṣa (Septum and Placenta)

Bījapuplika is the septum by which fruits are divided into chambers. It is also termed Bījapuṭa. It protects the seeds inside the fruit. It may be soft or hard. Bījapuṣa is a tissue, as found in some fruits, within which seeds remain embedded (cf. Placenta). Seeds will be described in the chapter - Bījāngasūtrīya. [Verse 19-20].

Bījabandhanaṁ (Arrangement of seeds in the fruit)

Seeds in the fruit may be arranged in five different ways, viz. Vṛntāntika, Puplikāntika, Puṣāntika, Sthālakāntika and Panktika. Vṛntāntika seeds are oriented towards the pedicel. Puplikāntika seeds are covered by the septa. Puṣāntika seeds are embedded within the placental tissue. Sthālakāntika seeds are placed on the thalamus. Panktika seeds are borne in rows along the margin. [Verse 21].

Abhrabīja phalaṁ

Some dehiscent fruits on ripening and drying burst open with a sound, throwing out the seeds at a distance, e.g. Kuruṇṭa (*Barleria cristata*). They are called Abhrabīja phalaṁ. [Verse 22].

Prakṛtito Phalagaṇa Vibhāga (Classification of fruits)

According to their nature, fruits may be divided into the following types : Tundika, Bījaka, Kurca and Āsphoṭaka.

Tundika type - The seed and the fruit are separate from one another. The seed can germinate independently without the presence of fruit pulp, e.g. Āmra (*Mangifera indica*), Jambu (*Syzygium* sp.), Dhātrī (*Phyllanthus emblica*), Badara (*Zizyphus jujuba*), Bimbi (*Coccinia indica*), and Kārpāsa (*Gossypium* sp.). [Verse 23 & 24].

Bījakagaṇīya Phalaṁ

The seed and fruit can no longer be distinguishable from one another and the entire structure germinates into the seedling, e.g. Śākavṛkṣa (*Tectona grandis*), Kāraikā (*Coriandrum* sp.), Kakubha (*Terminalia arjuna*), Sindhuvāra (*Vitex negundo*) Tṛṇadhānya (*Oryza sativa, Hordeum Vulgare*). [Verse 25].

Kurcagaṇīya Phalaṁ

Several fruits remain aggregated on the same thalamus looking like a brush (Kurcākāreṇa). The fruits are developed from the epigynous ovaries. On ripening they easily drop out of the thalamus. [Verse 26].

This type of fruit can again be divided into two groups, viz. Pucchika and Apucchika. In the Pucchika type the seeds develop a kind of hairy outgrowths which on mature seeds are spread like an umbrella over the thalamus and thus enabling the seeds to be carried away by wind, e.g. Somrāji (*Centratherum anthelminticum*). [Verse 27].

In Apucchika type seeds do not develop any such outgrowth, e.g. Keśarāja (*Eclipta alba*). [Verse 28].

Āsphoṭakagaṇīya Phalaṁ

These are the fruits which are developed from the flowers, usually, having two carpels and also sometimes with the staminal corona. The plants produce latex (Kṣīra vṛkṣa, Kṣīra means latex and Vṛkṣa means plant). These seeds have hairy outgrowth called Tulapuccha, which help them to be carried away by wind. The fruit wall splits open longitudinally along one side, e.g. Arka (*Calotropis gigantea*) and the other plants of the family i.e. Āsphoṭakagaṇīyaṁ (cf. Asclepiadaceae). [Verse 29].

Ākṛtito Phalagaṇa (Classification of fruits according to their shape)

Fruits mentioned above will now be classified according to their shape. Some are round, some being nearly round. Some fruits are long, some are flattened. Often they may be a combination of various shapes and are described as Miśrākṛti (Miśra means mixed and Ākṛti means shape). [Verse 30].

Summary

Fruits are distinguished by names also, e.g. Phalguphalaṁ, Kumbhaphalaṁ, Tripuṣam, Jālinī, Bhastrā, Śṛngī, Tulinī, Tripuṭa, Vṛntākī, Gucchaphalam, Gucchadantikā, Keśarakaphalaṁ, Koṣaphalaṁ, Śankuphalaṁ, Simbiphalaṁ and Bījakaphalaṁ. [Verse 31-32].

Now the characteristics of the above mentioned fruits will be described. [Verse 33-34].

Phalgu phalaṁ - The fruit is developed on a round, hollow thalamus. It will be described in detail in Vanaspati Kāṇḍa. [Verse 35].

Kumbha phalaṁ : The fruit is shaped like a pitcher (Kumbha means pitcher).

Tripuṣa phalaṁ : The fruit bears three placental columns.

Jālinī phalaṁ : The fruit when dried is reduced into a net like skeleton (e.g. *Luffa* sp.). [Verse 36].

Bhastrā phalaṁ : The fruit has a persistent calyx which resembles a leather air blower (Bhastrā' means a blower). [Verse 37].

Śṛngī phalaṁ : The fruit is shaped like a horn.

Tulinī phalaṁ : The fruit is filled with cotton or cotton like fibres.

Tripuṭa phalaṁ : The fruit has got three chambers (e.g. *Elettaria cardamomum; Operculina turpethum*). [Verse 38].

Vṛntākī : The fruit in which the pedicel is firmly attached to the calyx is called Vṛntākī (e.g. *Solanum melongena*) [Verse 39].

Guccha phalaṁ : Many fruits are aggregated on a common pedicel. They are called Guccha phalaṁ (e.g. *Vitis* sp.). [Verse 40].

Gucchadantikā phalaṁ : The fruitlets of Guccha phala are arranged like the row of teeth (e.g. *Musa* sp.). [Verse 41].

Keśaraka phalaṁ : The edible portion of the fruit consists of hair like outgrowths (Keśarākār), and the fruit is divided into several chambers (e.g. *Citrus* sp.) [Verse 42].

Koṣa phalaṁ : The fruit consists of many chambers with seeds inside, which are assembled on a solid thalamus. They are called Koṣa phalaṁ, (e.g. *Artocarpus heterophyllus*) [Verse 43].

Śanku phalaṁ : The fruit is shaped like a knife (e.g. *Holarrhena pubescenes*).

Samī phalaṁ : The fruit contains seeds borne in lateral rows. It dehisces longitudinally along the two lateral sides (e.g. *Albizzia* sp.). [Verse 44].

Bījaka phalaṁ : The fruit and seed are indistinguishable (e.g. *Tectona grandis* and *Vitex negundo*). [Verse 45].

Here briefly ends the classification of fruits. [Verse 46].

Rasataśca Phalagaṇa Vibhāga : (Classification of fruits according to their tastes)

Fruits may be divided into five classes according to the tastes of their juices viz. sweet, sour, pungent, astringent and bitter. The characteristic taste of a fruit is imparted by the juice that is predominating.

There are a great many different types of fruits. The nature of their juices vary according to the land types, and also depends on the elementary constituents which are governed by the Pāñcabhautik principles of Nature (e.g. Kṣiti, Ap, Teja, Marut and Vyom).

The taste of the juice also varies with the developmental stages of the fruit. [Verse 47].

Thus ends the Phalāngasūtrīyādhyāya, the sixth chapter of the Bījotpatikāṇḍa of the Vṛkṣāyurveda by Parāśara. [Verse 48].

बीजोत्पत्तिकाण्डः

अष्टाङ्गसूत्रीयाध्यायः ।

अथातोऽष्टाङ्गगणसूत्रीयमध्यायं व्याख्यास्याम इत्याह पराशरः ।।१।।

अस्मिन्नध्याये मूलत्वक्काण्डसारस्वरसनिर्यासस्नेहकण्टकाश्चेति वृक्षाणामष्टाङ्गगणविभागमुपदेक्ष्यामः ।।२।।

तत्र वृक्षाणां जीवनायनं मूलम् । तेन मूलेन भूम्याः पार्थिवादयः स्यन्दमानाः षड्रसा अभिगृह्णन्ते । यथा प्राणिनां मर्मच्छिन्ने निपातिता भवन्ति तथा वनस्पतेर्मूले छिन्ने स्कन्धशाखापत्रपुष्पफलादीनि प्रशोषमापदयन्ते ।।३।। तत्र मूलं ब्रध्नं शिफा पादश्चरणमित्यनर्थान्तरम्। अस्यैव स्वलक्षणैर्गणविभागश्चोपदेक्ष्यते ।।४।।

मूलानि हि प्रकृतिभिर्भिद्यमानानि चतुर्विधानि भवन्ति । तद् यथा काष्टिकञ्च कान्दिकञ्च ग्रन्थिकञ्च प्रतानिकञ्चेति।

तत्र मूलानि हि यानि काष्ठगर्भाणि तानि काष्टिकगणीयानि भवन्ति । काष्ठमन्तरेण मुलानि यानि पीवरपिण्डितानि रसबहुलानि तानि कान्दिकगणीयानि भवन्ति ।।५।।

प्रकृत्या कन्दश्च त्रिविधं भवति । पाट्टिकञ्च रुहकञ्चारुहकञ्चेति । तत्र पाट्टिकानि तु येषां मध्यस्थं बीजपत्रं बहुभिः पृथुलपट्टिकैश्च वेष्टितं भवति। अधस्तादस्यैव कन्दस्य मूलानि सञ्जायन्ते । एतदेव गुच्छमूलम्। अपि चेत कालप्राप्तानां प्रशुष्काणां येषां वल्लीगुल्मानां कान्दिकमूलानि न च प्रशुष्यन्तेऽपि च यथाकाले प्ररोहन्ति तानि रूहकाणि भवन्ति । एतेषु कन्देषु बीजत्वमभिजायते। अरूहकाणि तु न प्ररोहन्ति कन्दाङ्कुराभावात् । एव

प्रशोषमापदन्तेऽचिरेण ।।६।। ग्रन्थिकगणीयानि मूलानि हि ग्रन्थिबहुलानि भवन्ति ।।७।। यानि मूलानि भूमौ प्रतन्वन्ति तानि प्रतानिकगणीयनि भवन्ति ।।८।।

आकृतितो मूलानि खलु कानिचिद् दीर्घाणि नागदन्तप्रकाराणि कानिचित् पिण्डितवर्तुलानि कानिचित् शृङ्गभूतानि कानिचिद् गुच्छेन तन्तुसदृशनि भवन्ति ।।९।।

अथ मूलानि हि पाण्डुरक्तपीतनीलवर्णैस्तु विभज्यन्ते ।।१०।। मूलानि तु खलु यान्यवरोहन्ति शाखाकाण्डेषु तान्यवरोहगणीयनि भवन्ति ।।११।। इति मूलगणसूत्रीयः ।

अथ त्वग्गणसूत्रीयः ।

इह खलु त्वगावरणं वल्कलञ्चेत्यनर्थान्तरम् ।।१२।। तत्र वृत्राणां सर्वस्त्रोतांसि स्यन्दनी सिरा जालानि निवद्धानि भवन्ति । एतेषां स्त्रोतसां रसवहस्त्रोतोभिः स्यन्दनीर्भिमूलात् पृथिव्याः स्यन्दमाना रसाः संप्लवन्ते वृक्षाङ्गेषु । तस्माद् वृक्षाः सन्तर्पयन्ते वर्धन्ते च । एव सिराभिश्चोपसर्पयन्ति अपसर्पयन्ति च भावाभावौ । व्युह्यमानत्वात् क्षतसंरोहणत्वाच्च वृक्षशरीरे वायुरेव चानुमतव्यम् ।।१३।। महीरुहाणां काण्डैकदेशे समन्ततो वल्कले छिन्ने काण्डशाखापत्रादीनि प्रशोषमापद्यन्ते यतो मूलगतानां स्त्रोतसां वल्कलेन सह संयोगाभावात् । तत्र सिराः खलु वायुवाहिन्यः स्त्रोतांसि भवन्ति। स्यन्दन्यस्तु खलु रसवाहिन्यश्च ।।१४।।

अथास्य वल्कलस्य खलु गणविभागो वर्णतः सितपाटलहरितपीतकपिशाश्चेति ।।१५।।

अतो वल्कलं प्रकृतितश्चांशुकपत्रकपाटलसारमृदुलखरकाण्टिकानीति विद्यात् ।।१६।। तत्र वल्कालानि हि यानि तन्तु बहुलानि तान्यंशुकगणीयानि भवन्ति । वारगणीयमित्येके ।।१७।। यानि तु वल्कलानि पत्रकेणोपर्युपरि निसृष्टानि तानि पत्रकगणीयानि भवन्ति। यानि खलु पाटलेनोपर्युपरि निसृष्टानि तानि पाटलगणीयनि ।।१८।। यानि वल्कलानि खलु सारभावत्वेन चोपलक्षणीयानि तानि हि सारगणीयानि भवन्ति ।।१९।। यानि वल्कलानि तु बहुलरसयुक्तत्वेन

मृदुनि भवन्ति तानि मृदुलगणीयानि ।।२०।। यानि वल्कलानि खलु वैशद्यगुणत्वेन कठोराणि तानि खरगणीयानि भवन्ति ।।२१।। यानि वल्कलानि कण्टकयुक्तानि तानि कण्टकगणीयानि भवन्ति इति त्वगगणसूत्रीयः ।।२२।।

अथ काण्डगणसूत्रीयः ।

खलु काण्डः प्रकाण्डः स्कन्ध इत्यनर्थान्तरम् ।।२३।। काण्डस्तु मूलानामूर्ध्वं तिष्ठति । एव शाखानामाश्रयः ।।२४।। काण्डात् शाखाश्च तथा शाखाभ्यः सम्भवन्ति पुष्पपल्लवाश्च ।।२५।।

द्विविधः खलुकाण्डः पत्रप्रभवः पुष्पप्रभवश्च । यस्तु पत्रेभ्यो वर्तते स च पत्रप्रभवः । पत्रप्रभवकाण्डस्तु स्थाणुकाण्ड इति विज्ञायते। पुष्पप्रभवस्त्वक्षकाण्ड इति विज्ञायते स च पट्टिकवेष्टितेन बल्वजानाञ्चैव कण्डुलानां पुष्पकाले सम्भवति। तस्मादक्षमञ्जरिकामाचक्षते एव फलपाकान्तश्च ।।२६।। अतः पत्रप्रभवकाण्डाः प्रकृतिर्भिभिद्यमानाः षड् विधास्तु यथा दार्विककीलकास्त्रकनिःसारपार्विकसौषिराश्चेति । तत्र दार्विकः काष्ठभूतश्च । कीलककाण्डस्तु काष्टभूतैस्तीक्ष्णकीलैः संकीर्णो भवति। त्रिभिर्वा चतुर्भिरस्त्रकैरस्त्रकं विद्यात् । साररहितेन निःसारं विद्यात् । पर्वभिर्बहुभिर्युक्तैः पार्विकं विद्यात् । सुषिरेण सौषिरं विद्यात् ।।२७।।

प्रकृतिविशेषण वृक्षाणां काण्डाच्छाखाः सम्भवन्ति । यथाभूतं सम्भवति तदुपदेक्ष्यामः ।।२८।।

खलु पत्रकुक्षौ शाखानां सम्भवो भवति । केषाञ्चिद् वृक्षाणां पृथक्श्रेण्या केषाञ्चित पक्षश्रेण्या केषाञ्चिदरकसंस्थानेन केशाञ्चित् साङ्कर्येण वा शाखा सम्भवति । यत्र यत्र यद् यत् संस्थानं भवति तत्र तत्र तदधिकृत्य संज्ञायते ।।२९।।

भवति चात्र ।

पत्रेभ्यो वर्तते काण्डः काण्डादेव शाखा स्मृता ।
पत्रकुक्षौ भवेच्छाखा शाखायां पुष्पपल्लवौ ।।३०।।
इति काण्डगणसूत्रीयः ।

अथ सारगणसूत्रीयः ।

पादपानां वल्कलाभ्यन्तरे यत् कठिनं स सारः । यथा देहिनोऽस्थि सारैदेहं धारयन्ति तथा महीरुहा अभ्यन्तरगतसारेण तिष्ठन्ते । स च सारः पर्यायान्तरेणेपचीयते कालपरिनामाच्च । केषाञ्चिद् वृक्षाणां सारो वर्णान्तरेण परिवर्तते कालप्रकर्षाच्च। तस्मांद् वृक्षविशेषेण सारस्तु सितारुणरक्तपीतधुम्रकृष्णाश्चेति वर्णतो विभज्यते ।।३१।। पुनश्च सारः स्नेहगन्धाभ्यां द्विविधः । तत्र स्नेहयुक्तेन स्नेहसार एव गन्धयुक्तेन गन्धसार इति विज्ञायते ।।३२।। इति सारगणसूत्रीयः।

अथरसगणसूत्रीयः

अव्यक्तभावतन्मात्रात् खलु वृक्षशरीरेषु व्यक्तिभावरसा अभिनिर्वर्तन्ते। तेषां रसानां योनिरुदकम् । ते हि पाञ्चभौतिकगुणयोगात् स्थावराणां शरीरमभिप्रीणयन्ति । तस्माद्धिवृक्षशरीराणि खल्वन्तरीक्षवाय्वग्नितोयभूमीनां विकारहेतुभूतानि भवन्ति । तत्र वृक्षाङ्गेषु येभ्यो येभ्यो भावेभ्यो महाभूतविकारा मन्यमाना भवन्ति तदनु व्याख्यास्यामः ।।३३।।

तत्राकाशात्मकं शब्दं लाघवं सौक्ष्म्यं सौषिर्यम् ।।३४।।

वाय्वात्मकं स्पर्शः रौक्ष्य व्युहनं चेष्टाश्च ।।३५।।

अग्न्यात्मकं वर्णमौषण्यञ्च ।।३६।। अबात्मकं रसः शैत्यं मार्दवः स्नेहञ्च ।।३७।। क्षित्यात्मकं गन्धः गौरवं स्थैर्यं मूर्तिश्च एव स्थावराणां पाञ्चभौतिकं शरीरमुपदिशन्ति कुशलाः ।।३८।।

अतो वृक्षशरीरे पञ्चभूतगुणसमन्विता जिह्वावैषयिकाः षड्रसा उक्ताः मधुराम्ललवणकटुतिक्तकषायाश्चेति ।।३९।। अपि चेदेकरसः कस्मिंश्चिद् वृक्षे नास्ति तत्रापि य्रो रसः प्राधान्येनोपलभ्यते तमभिप्रेत्य रसगणमुपदिशन्ति रसज्ञाः ।।४०।। एषां षण्णां रसानामापगुणातिरेकान्मधुरो रसः ।।४१।। सलिलाग्निगुणातिरेकादम्लः ।।४२।। क्षित्यग्निगुणातिरेकाल्लवणः ।।४३।। वाय्वाग्निगुणातिरेकात् कटुकः ।।४४।। वाय्वाकाशगुणातिरेकात्तिक्तः । पवनभूम्यतिरेकात् कषायः । एव पञ्चभूतगुणसमन्विताः षड् रसा वृक्षाणां पाञ्चभौतिकशरीरमभिप्रीणयन्ति ।।४५।। इति रसगणसूत्रीयः ।

अथ निर्यासगणसूत्रीयः ।

वृक्षवल्लीगुल्मानां शाखापल्लवच्छिन्ने काण्डे शस्त्रक्षते वा तत्र यच्च निःस्रवति तद्धि निर्यासम् ।।४६।। प्रकृतिभेदेन निर्यासानि परिप्लवफेनसान्द्रक्षीरमोचजतुक संज्ञकानि स्युः । तत्र परिप्लवन्तु सलिलसंङ्काशम् । फेनस्तु फेनसङ्काशम् । सान्द्रञ्च घनीभूतं कललवदच्छम् । क्षीरनिर्यासन्तु क्षीरसङ्काशम् । मोचनिर्यासं पिच्छिलं साटकमस्नेहश्च । जतुकन्तु बहुस्नेहान्वितंपिण्डकं गन्धोपेतं बह्निदहनक्षमञ्च ।।४७।। इति निर्यासगणसूत्रीयः ।।

अथ स्नेहगणसूत्रीयः ।

इह खलु केषाञ्चिदुद्भिदानां पत्रपुष्पसारबीजादीनि स्नेहयुक्तानि भवन्ति ।।४८।। तेषामुद्भिदानां कस्यचित् पत्रे कस्यचित् पुष्पे कस्यचिद् बीजे कस्यचिन्निर्यासे कस्यचित् सारे स्नेहोऽस्ति । येषु येषु चाङ्गेषु स्नेहोऽस्ति तत्तदङ्गमधिकृत्य तत्तत् संज्ञायते ।।४९।।

स्नेहस्तु प्रकृत्या द्विविधो भवति । विरलं सान्द्रञ्च । तत्र विरलं यद्धि शीतेन तनुभावं न जहाति । सान्द्रन्तु सोमगुणत्वात् शीतेन घनीभूतं भवति चोष्णेन द्रवयति ।।५०।। इति स्नेहगणसूत्रीयः ।

अथ कण्टकगणसूत्रीयः ।

केषाञ्चिद् वृक्षाणां वल्कले पत्रे वा शङ्कुसङ्काशं तीक्ष्णाग्रं यदङ्गं दृश्यते स हि कण्टकः ।।५१।। कण्टकस्तु प्रकृत्या त्रिविधो भवति। ऋजुवक्रकुटाश्चेति । तत्र ऋजुकण्टकस्त्वनताग्रो वक्रोऽवनताग्रः कुटकण्टकस्तु स्थूलकण्टक एव शिखराकारश्च । वृक्ष एव कण्टकेन संज्ञायते ।।५२।। इति कण्टकगणसूत्रीयः ।

इति पराशरकृते वृक्षायुर्वेदे बीजोत्पत्तिकाण्डेऽष्टाङ्गसूत्रीयो नाम सप्तमोऽध्यायः ।।५३।।

Bījotpatti Kāṇḍa
Chapter - VII
Aṣṭāngasūtrīyādhyāya

Parāśāra said : I will now delineate the Aṣṭāngasūtrīyādhyāya [Verse 1]. In this chapter eight plant parts, (other than shoots, leaves, flowers etc.) have been described. They are termed as Mūlam - the root; Twak - the bark; Kāṇḍa - the stem or trunk; Sāra - the heart - wood; Swarasa - the sap (the fluid that circulates in a plant); Niryāsa - the exudate like oleoresinous substances (incense, gum, etc.); Sneha - oleaginous substances (essential oil in the heart-wood of *Santalum album*); Kaṇṭaka - spine and prickle. They are collectively called 'Aṣṭānga'. [Verse 2].

The root is the main life-spring (Jīvanāyanaṁ) of the plant. (Note : Hence the word 'Mūlam' is etymologically quite appropriate). The root absorbs and transports (Syandamānah) the saps of six different tastes from the soil.

Just as an injury to the heart of an animal causes its eventual death, similarly when the root of a plant is damaged the plant dies and its trunks, branches, leaves flowers dry up eventually. [Verse 3].

Synonyms of root

The synonyms of root are - Mūlam, Bradhnaṁ, Śiphā, Pāda and Caraṇa.

Parāśara said : Roots are classified according to their characteristics. [Verse 4].

The four distinguishing characters of roots

According to their nature roots are of four distinct types. These are known as Kāṣṭika, Kāndika, Granthika, and Pratānika.

Kāṣṭika : It is called Kāṣṭika root when its core is woody (Kāṣṭa - garbha).

Kāndika : Roots which have no woody core, but have thick fleshy substances (Pīvara) containing enough sap (Rasavahulāni') belong to the type 'Kāndika gaṇa'. [Verse 5].

Kāndika root (bulb, rhizome) is again subdivided into three classes, viz. Pāṭṭika Ruhaka, Aruhaka.

Pāṭṭka Kanda : In the Pāṭṭika type, the primary leaf is surrounded by numerous fleshy layers (Pṛthulapaṭṭika) and have got bunches or rootlets below the bulbs called Gucchamūlam (cf. bulb of *Allium cepa*).

Ruhaka Kanda : In some shrubs and creepers the underground rhizomes, tubers or bulbs (Kanda) never dry up and they sprout at a particular time of the year. These organs have the inherent property of sprouting.

Aruhaka Kanda : Though some plants have 'Kanda-Mūlam', still they are not capable of being propagatged by their Kanda. (Best examples are - *Raphnus sativus* Sans. Mūlaka, and *Daucus carota,* Sans. Gṛñjanaka). These Kandas simply dry up naturally.

Granthika

Some plants have knotty roots: (e.g. *Piper longum, Piper chaba,* Sans. Pippali and Cavikā respectively). [Verse 7].

Pratānika

Some plants have roots that spread like creeper. They are called Pratānika Mūlam, (Runners of *Colocasia esculenta*). [Verse 8].

Ākṛtito Mūlam (Forms of roots)

Roots have got different shapes. In some class of plants they look like elephant's tusk (e.g. *Alocasia indica* Sans. Mānaka); sometimes they appear solid and round; often they may look like horns. In many plants they look like bunch of threads. [Verse 9].

Varṇata Mūlām (Colour of roots)

The colour of a root also varies from plant to plant, and may appear in different shades like pale white, dirty yellow, red or blue [Verse 10].

Avaroha Mūlaṁ

In some trees, roots orginate from their trunks and branches, (cf. Adventitious roots, e.g. *Ficus bengalensis* Sans. Vaṭa.). They are called Avaroha Mūlam. [Verse 11].

Here ends the Mūlagaṇasūtrīya.

Twakgaṇasūtrīya : (Vascular system of the plants)

Twak and Valkal are synonymous (denoting bark). [Verse 12]. The plant has a circulatory system consisting of 'Syandanī' and Sirā. Of these, Syandanī performs the function of transporting (The Sanskrit root: 'Syandan' means transport) the elementary fluid (Pāñcabhautik rasa) from the earth with the help of root. Thus the plant receives nourishment, and grows. Through Sirā the fluid circulates both in the inward and outward directions (... 'eva upasarpayanti apasarpayanti ca bhāvābhāvau'). It is presumed (Anumatavyaṁ) that from the ability of having general functional capacity (Vyuhamānatwāt) and also of the healing of injuries, a plant, necessarily, has the presence of 'Vāyu' (as in the human body; 'Vāyu' or 'Vāta' is equivalent to the nervous system - that feels and performs).* [Verse 13].

The bark covering the stem entirely, when cut through causes drying up of all the plant organs ('Praśoṣamāpadyante') because the circulatory system to the root is cut off. Sirā circulates air ('Vāyu' or 'Vāta'); and Synadanī carries watery fluid or sap. The bark will be now classified according to colour which may be white, pink, green, yellow or variegated. [Verse 14-15].

Prakṛtito Valkala (Physical character of bark)

Parāśara said : Bark of different trees varies variously. Accordingly it is called 'Angśuka'. 'Patraka', 'Pāṭala', 'Sāra', 'Mṛdula', 'Khara' and 'Kāṇṭika'. [Verse 16].

Angśuka bark : A bark is called 'Angśuka' when it is composed of thread-like fibres (e.g. *Corchorus* sp. *Urena lobata* etc.). It is also called Vāragaṇīya. [Verse 17].

Patraka bark : 'Patraka' bark consists of papery layers, (e.g. 'Bhojpatra' *Betula bhojpattra*). [Verse 18].

Pāṭala bark : A bark is called 'Pāṭala' when several layers are placed successively (e.g. *Stereospermum suaveolens* Sans. Pāṭala - one of the ten herbs of 'Daśamūla', of Āyurveda).

Sāragaṇīya bark : A bark is called 'Sāra-gaṇīya' when it is substantially rich (as in the case of bamboo i.e. *Bambusa* sp.) [Verse 19].

---

*. This view of Parāśara is also tenable from the fact that he earlier said that plants were conscious of 'Sukha and Duhkha' ('bhavedantah sukha dukha samanwitah').
The nervous mechanism of plants by J.C. Bose (1926).

Mṛdulagaṇīya bark : A bark containing excessive sap and thus soft is called 'Mṛdula-gaṇīya' (e.g. *Moringa pterygosperma* Sans. Sobhanjana). [Verse 20].

Kharagaṇīya bark : When a bark has predominantly rough surface it is called 'Khaṛa-gaṇīya' bark ('Khara' means rough). [Verse 21].

Kaṇtakagaṇīya bark : A bark is called 'Kaṇṭakagaṇīya' when its surface is extensively covered by spines. (e.g. *Salmalia malabarica*). [Verse 22].

Kāṇḍagaṇasūtrīya

Kāṇḍa (Stem) has several synonyms : viz. 'Kāṇḍa', 'Prakāṇḍa' and Skandha. [Verse 23].

The 'Kāṇḍa' or the stem is formed above the root system and it supports the branches. [Verse 24].

The stem produces branches ('Śākhā) which bear flowers. [Verse 25].

Kāṇḍa is classified into two groups, viz. 'Patra-prabhāva' and 'Puṣpa-prabhāva'. Patra-prabhāva kāṇḍa develops from leaves. Patra-prabhāva kāṇḍa is persistent. Puṣpa-prabhāva kāṇḍa is known as 'Akṣakāṇḍa'. In this case the stem is ensheathed by a spathe which develops at the time of flowering as found in the families of grasses and esculents. It dries up at the time of maturity of the fruits. [Verse 26].

Patra-prabhāva kāṇḍa is classified into six categories, viz. 'Dārvika', 'Kīlaka', 'Asraka', 'Nihsāra', 'Pārvika', and 'Sauṣirā'.

Dārvika kāṇḍa : It is mostly composed of woody tissue.

Kīlaka kāṇḍa : It is woody and covered by spines.

Asraka kāṇḍa : A triangular or quadrangular shaped stem is called an Asraka kāṇḍa (example-triangular : *Typha elephantina,* quadrangular : *Cissus quadrangularis*).

Nihsāra kāṇḍa : A stem deficient in substance is called Nihsāra kāṇḍa (e.g. *Ipomoea reptans*).

Pārvika kāṇḍa : A stem has numerous segments (as found in *Bambusa bambos, Saccharum officinarum* etc.).

Sauṣira kāṇḍa : A stem which is hollow (as in *Bambusa* sp. and grassy plants) is called 'Sauṣirā kāṇḍa'. [Verse 27].

Śākhā - the branches

Parāśara said : By nature, branches (of plants) develop from trunks. Generally the branches come out of leaf axils ('Patra kukṣau').

In some plants, branches form at alternate sites (Pṛthak Śreṇyā). In other cases they develop at opposite sites. Then again they develop in the fashion of wheel axles ('Arakasamsthānena'). Sometimes, their arrangement is quite irregualr; they develop here and there on the trunk. They are termed therefore according to the positions from which they originate. [Verse 28-29].

Summary

Leaves nourish the trunk, and the trunk nourishes the branches. Branches come out of the leaf axils. And branches in turn, bear leaves and flowers. [Verse 30]. Here ends the chapter on Kāṇḍa gaṇa sūtra.

Sāragaṇasūtrīya : It is the subject on the heart - wood. In a tree the hard portion inside the bark is called 'Sāra' - the heart-wood. Just as in the case of an animal the bony skeleton provides support to the body, similarly in a plant the heart-wood enables it to stand erect. The heart-wood is developed through years, by successive layers ('Paryāyāntareṇa Upacīyate'). In case of some trees the heart-wood acquires different kinds of colours in course of time. Hence trees may have different colours. It may be red, grey, golden, yellow, black, white and varieties of admixture. [Verse 31].

Again, the heart-wood may be classified into two kinds: having oleaginous principle ('Snehayuktena') and aromatic principle ('Gandhasāra'). [Verse 32].

Rasagaṇasūtrīya

Parāśara said : I will now deal with the subject of 'Rasa'. This 'Rasa' (watery juice) is to be conceived (according to the 'Sāmkhya Darśana' of Indian philosophy) right from the basic invisible matter ('Avyaktabhāvatanmātrāt') i.e. Tanmātra (cf. particualte matter). Now this Rasa as perceptibly present (Vyaktibhāva Rasa) in the plant life has its origin in water (Yonirudakam). It nourishes the plant organs with all the derivatives of five 'Pāñcamahābhautik' elementary matters viz. 'Kṣiti', 'Āp', 'Teja', 'Vāyu' and 'Ākāś. Now the various qualities and characteristies of the plants as attributable to their respective 'Mahābhūta' are described below. [Verse 33].

The cause of lightness, fineness, hollowness, etc. in a plant body is due to the overriding influence of 'Ākāś' mahābhūta, one of the five elements of 'Pañcamahābhūta'. Similarly, the cause of touch, roughness, and control or regulation (Vyuhan) of different functions are due to the element of 'Vāyu' (Air). [Verse 34].

'Agni' or 'Teja', the fiery element is responsible for creating all sorts of colorific principles (in a plant body) and heat. [Verse 35].

The 'Āp' or the aqueous element is the cause for maintaining the coolness ('Śaitya'), softness ('Mṛdutwa') and oleaginous property in a plant. Lastly, the 'Kṣiti' or the earthy element is responsible for creating aromatic principles (as in flowers and heart-wood, etc.), heaviness and distinct image (gandhah gauravaṁ sthairyam mūrtiśca). Thus a scholar interprets the complete form of a tree (Sthāvarāṇam) in terms of the five elementary principles of 'Pañcabhūta'. [Verse 36-38].

I will now discuss the six 'Rasas' present in the plant body, pertaining to the taste, and endowed with the quality of five elementary principles (Pañcamahābhūta). There are six different kinds of tastes or 'Rasa', viz. sweet, sour, saline, pungent, bitter and astringent. Rarely in a plant only a single taste is present. However; the principal constituent of Rasa that has an overtone over the others is more apparent and accoridngly the schoalrs classify the plant juice into different categories. [Versse 39-40]. Out of the six tastes, the sweet taste (Madhura Rasa) has more of the elementary property of 'Āp' guṇa. [Verse 41]. Sour tasting substances (Amla Rasa) have the dominating influence of aqueous and igneous property of the two elementary mahābhūtas, viz. 'Āp' and 'Agni' or 'Teja'. Similarly, salinity has the predominating influence of the two elementary mahābhūtas , viz. 'Kṣiti' and 'Agni' or 'Teja'. [Verse 42-43].

Note : The relevancy of such philosophical contention, by Parāśara, in the evolution of Lavana Rasa out of the collocation of 'Kṣiti' and 'Agni' is quite tenable. And to understand the inner significance of this contention one has just to conceive as to how a salt is formed by the reaction of a base or basic substance with acidic substance against the perspective of Chemistry. Needless to say a base is an earthy and an acid is a fiery substance.

Due to the predominance of 'Vāyu' (airy) and 'Agni' (fiery) property, the plant body acquires the pungent Rasa (Kaṭu guṇa). [Verse 44].

In the same way due to the dominating influence of 'Vāyu' and 'Ākāś' elements, bitter tasting Rasa develops in a plant body.

Note : The alkaloid quinine of *Cinchona* plant when used as an antimalarial drug, the malaria patient invariably has the feeling of giddiness or dizziness as a side effect. Now considering the above view if one tries to understand this property of quinine i.e. giddiness, it becomes easy to realise the involvement of 'Ākāś' and 'Vāyu' elements of the Pañcamāhbhūta' in the formation of bitter tasting Quinine in *Cinchona* plant. In fact Quinine is an intensely bitter substance.

Lastly, due to the predominance of 'Pavana' or 'Vāyu' along with 'Bhūmi', i.e. airy and earthy elements of the complements of the 'Pañcamāhbhūta', 'Kaṣāya Rasa' that is astringency develops in a plant body.

Thus the six different kinds of 'Rasas' endowed with Pāñcabhautic principles, nourish the plant body. [Verse 45].

Here ends the discourse on the Rasagaṇasūtrīya.

Niryāsagaṇasutrīya

Now the subject on the formation of Exudation (Niryāsa) in plant begins here. In trees, creepers, shrubs when their branches, leaves etc. are cut off or their trunk is damaged by a sharp instrument, a kind of exudation comes out. It is called 'Niryāsa'. [Verse 46].

Exudation, according to their nature has different physical properties. Some may be free flowing watery fluid (Pariplava salilasaṁkāśaṁ). Others may be frothy (Phena); slimy (Sāndra); again others may be milky juice i.e. latex (Kṣira e.g. *Calotropis gigantea*). Sometimes they even may be sticky - ('Moca', e.g. *Salmalia malabarica)* or oleo-resinous (Jatuka, e.g. *Commiphora mukul*). The former is devoid of oil, while the latter is an inflammable substance. [Verse 47].

Snehagaṇasutrīya

Now begins the subject on the formation of oleaginous principle (Sneha). [Verse 48]. The leaf, flower, heart - wood (Sāra), seeds, etc. and also the exudates (Niryāsa) in some plants contain oleaginous substances - 'Sneha'. These oleaginous properties are named according to their respective origin. Oleaginous substances are of two kinds, viz. liquid and solid ('Viralam Sāndranca'). The oleaginous substances that do not congeal by exposure to cold are called 'Virala sneha'. Whereas the other variety that gets solidified on exposure to cold because of 'Soma guṇa' are called 'Sāndra sneha'. [Verse 50].

Kaṇṭakagaṇasūtrīya

Parāśara said : I will now discuss about spines and prickles. The knife like sharp projections found on the bark and leaves in some trees are known as Kaṇṭaka - spines, prickles. According to their nature, they are of three types, viz. 'Ṛju', 'Vakra', 'Kuṭa'. Of these, 'Rju' is straight, unbent. 'Vakra' type is bent downward. 'Kuṭa' is broad based and is pointed like a peak. Thus trees can be specified by their spines.

Here ends the seventh chapter, the Aṣṭāngasūtrīya Adhyāya of the Bījotpatti Kāṇḍa of the Vṛkṣāyurveda by Parāśara [Verse 51-53].

बीजोत्पत्तिकाण्डः

द्विगणीयोऽध्यायः ।

अथातो बीजाङ्कुरे द्विगणीयमध्यायं व्याख्यास्याम इत्याह पराशरः ।।१।।

इह खलु बीजन्तु वृक्षाणां सम्भवकारणं भवति । विशेषतो बीजानां स्वभाव उद्भेदनञ्च । यस्मात् प्ररोहः सम्भवति तच्च बीजम् ।।२।। खलु कीखोस-बीजमातृकाबीजपत्रमातृकाच्छदाश्चेति बीजाङ्गानि भवन्ति ।।३।। तत्र बीजस्यावरणं कीखोसम् । तच्च काष्टिककर्परमृदुलभेदेन त्रिविधं विद्यात् । तत्र काष्टिकं काष्ठोपमम् । कर्परं खर्परवद् भङ्गुरम् । मृदुत्वेन मृदुलम् ।।४।।

बीजमातृका तु बीजशस्यम् । बीजपत्रन्तु बीजमातृकाया मध्यस्थमादिपत्रञ्च । मातृकाच्छदस्तु तनुपत्रकवत् मातृकाच्छादनञ्च कञ्चुकमित्याचक्षते ।।५।।

बीजन्तु प्रकृत्या द्विविधं भवति । एकमातृकं द्विमातृकञ्च । तत्रैकपत्रप्ररोहाणां वृक्षाणां बीजमेकमातृकं भवति । द्विपत्रप्ररोहाणान्तु द्विमातृकञ्च ।।६।।

खलु न्यूनातिरेकविशेषान्महाभूतानामेव वृक्षाणां फलबीजानां नानावर्णाकृतिविशेषाश्च भवन्ति । तत्र वर्णतो बीजानि तु श्वेतरक्तपीतनीलादीनि नानाविधानि मिश्रवर्णानि वा भवन्ति ।।७।।

आकृतितश्च बीजानि दीर्घवर्तुलावर्तुलपृथुकपुङ्खपुच्छिकस्थूलसूक्ष्मादीनि नानाविधानि । कस्यचिद् बीजस्य धारावत् कर्णः दृश्यते । तत्तु धाराबीजमाचक्षते ।।८।।

बीजानि तु कालप्राप्ते भूमौ प्ररोहन्ति । वृक्षरुहाणां बीजन्तु वृक्षकाण्डे प्ररोहन्ति।।९।।

केषाञ्चिद् गुल्मक्षुपाणां बीजं नास्ति तेषां पुष्पफलाभावादवकेशी विज्ञायते ।।१०।। केषाञ्चिदवकेशिनां पत्रपृष्ठे बीजस्वभावः पत्राङ्कुरः सञ्जायते। एतेन पत्राङ्कुरेण ते गुल्मक्षुपादयः समुद्भवन्ति ।।११।।

केचिद् वृक्षवल्लीगुल्माश्छिन्नकाण्डे प्ररोहन्ति । तेषामपि काण्डगात्रे काण्डाङ्कुरः सञ्जायते । एव क्वचित् कन्दे सञ्जातः कन्दाङ्कुरेण क्वचिन्मूले सञ्जातो मूलाङ्कुरेण प्ररोहयते । एते अङ्कुराश्च बीजस्वभावा उद्भेदत्वाच्च बीजगणीये व्याख्यातः ।।१२।।

केषाञ्चित् सफलपुष्पाणां बीजाधारे बीजं पूर्णाङ्गं न जायते । तेषां पुष्पाणां बीजपोष्ट्ररसादयश्चान्यतमेषु खल्वङ्गेषु प्रभुतेन संप्लवत्वाद् बीजाधारे बीजं न जायते न वा परिपुष्टञ्च (Lacuna exists), बीजपोष्ट्ररसाभावात्। एतेषां पुष्प वृक्षाणां बीजपोष्ट्ररसादयो यस्मिन् यस्मिन्नङ्गे संप्लवते प्राचुर्येण तस्मिस्तस्मिन्नङ्गे बीजभावमुत्पन्नं भवति । तस्मात् कस्यचित् कन्दे कस्यचित् काण्डे कस्यचित् पत्रे बीजभावं सञ्जायते । एतस्मादङ्गादेव प्ररोहत्वाद् वृक्षास्तत्तदङ्गमधिकृत्य ते संज्ञायन्ते । यथा कन्दरुहः काण्डरुहः पत्ररुहश्चेति । एतदेत्र विकृतेन भवति । कश्चिदाह स्वभावेन भवति । न तत् सम्यक् यतः प्रथमबीजाभावादिति बीजगणसूत्रीयः ।।१३।।

अथ प्ररोहगणीयम्

इह खलु प्ररोह उद्भेदोऽङ्कुर इत्यनर्थान्तरम् ।।१४।।

अङ्गानि खल्वस्य प्ररोहस्य तु बीजमातृका बीजपत्रमूलानीति भवन्ति ।।१५।। यदा भूमौ निपतितं बीजमुच्छजत्वं मृदुत्वं मूलभावञ्च याति तदाङ्कुरोऽभिनिर्वर्तते । ततो बीजमातृका व्यक्ताभवति। क्वचिद् बीजपत्रञ्च ।।१६।। अङ्कुरनिर्वृत्ते बीजमातृकाया रसः संप्लवते प्ररोहाङ्गेपु । तेनैव रसेन प्ररोहः स्निह्यते वर्धते च यावन्मूलं न स्वतन्त्रवृत्तिः स्यात् ।।१७।।

यदा प्ररोहः स्वातन्त्र्येण भूम्याः पार्थिवरसं गृह्णाति तदा बीजमातृका प्रशोषमापद्यते । ततो मूलेभ्यः पार्थिवरसः स्यन्दनीभिः संप्लवते पत्रेषु। तत्र रञ्जकेन पच्यमानात् यन्मलं निर्गच्छति तदुष्मा भवति । ततः पुष्यन्ति

पच्यमानात् सूक्ष्मरसादस्य सर्वाङ्गाणि । तस्मात् प्ररोहः कालपरिणामात् सर्वाङ्गेण परिपूर्णः भवति ।।१८।।

खलु बीजमातृकाभेदेन प्ररोहो द्विविधो भवति । एकमातृकोद्विमातृकश्च । तत्रैकमातृकप्ररोहाणां प्ररोहे एकपत्रं भवति । द्विमातृकप्ररोहाणान्तु द्विपत्रञ्च । प्ररोहस्तु पुनः प्रकृत्या त्रिविधो भवति । अव्यक्तमातृकश्च व्यक्तमातृकश्च विदलश्चेति । तत्राव्यक्तमातृकप्ररोहस्य कठोरकीखोसेन मातृका संवीता भवति। व्यक्तमातृकस्य तु मृदुलकीखोसत्वाद् मातृका व्यक्ता भवति । विदलस्य बीजे बीजमातृका कुञ्चिता भवति । ततः प्ररोहोद्गमे बीजमातृका विशिष्टदलेन व्यक्ता भवति ।।१९।।

नामतश्च प्ररोहः पञ्चविधो भवति । तद् यथा बीजाङ्कुरश्च पत्राङ्कुरश्च काण्डाङ्कुरश्च कन्दाङ्कुरश्च मूलाङ्कुरश्चेति । तत्र बीजेभ्यो ये चाङ्कुराः सम्भवन्ति ते बीजाङ्कुराः । एव पत्रकाण्डकन्दमूलेभ्यः सम्भवाश्चाङ्कुरास्तेन तेनाङ्गेन विज्ञायन्ते ।।२०।।

वृक्षवल्लीगुल्माश्च येन येनाङ्गेन प्ररोहन्ति तेन तेनाङ्गेन स स गणः संज्ञायते । यथा बीजेभ्यो बीजरुहाः पत्रेभ्यः पत्ररुहाः काण्डेभ्यः काण्डरुहाः कन्देभ्यः कन्दरुहा मूलेभ्योमूलरुहाश्चेति ।।२९।। बीजानां प्ररोहे भूम्यपवातातपादयो हेतुभूताश्च भवन्ति । तैरेव प्ररोह आप्यायते वर्धते च ।।२२।।

अतः पृथिवी धारणे वर्धणे, आपः संश्लेषणे, वात उद्भेदने, आतपो दहने रञ्जने एव कारणं भवति ।।२३।।

प्ररोहस्य यानि यानि खल्वाङ्गानि पाञ्चभौतिकानि तान्यनुव्याख्यास्यामः । तद् यथा बीजमातृका मूलं गन्धश्चेति पृथिव्यात्मकानि । अबात्मकानि तु स्यन्दनीरसस्नेहाश्चेति । औषण्यं वर्णः पत्रञ्चेति तैजसात्मकानि । सिराजालतन्तवश्चेति वाय्वात्मकानि । आकाशात्मकानि तु छिद्राणि स्युः ।।२४।।

यदि चेत् कालवैगुण्यात् पृथिव्यादीनामन्यतमस्य वा हीनातियोगमिथ्यायोगाच्च बीजे यस्य यस्याङ्गस्य बीजभाव उपतप्तो भवति तदा प्ररोहस्य तस्य तस्याङ्गस्य विकृतिरुप जायते ।।२५।।

भवन्ति चात्र ।

कललाज्जायते पिण्डं पिण्डाद् बीजत्वमुद्भवम् ।
बीजाङ्कुराद् भवेन्मुलं मूलात् पर्णं समुद्भवम् ।।२६।।
पर्णात्मको भवेत् काण्डः काण्डाच्छाखा प्रजायते ।
शाखायाश्च भवेत् पुष्पं पुष्पात् फलं ततो बीजम् ।।२७।।
एष क्रमो बीजोत्पत्तेनिसर्गाद् भुवि जायते ।
इति प्रोक्तौ बीजाङ्कुरौ द्विगणे यस्य यद्भवेत् ।।२८।।

तत्र श्लोकाः ।

सूत्रं यच्च बीजात्पत्तेर्भूमिवनानि यानि च ।
वृक्षाङ्गपत्रगणश्च पुष्पफलौ पृथक् पृथक् ।।
मूलत्वक्काण्डसाराश्च निर्यासौ स्नेहकण्टकौ ।
इत्यष्टगणसंग्रहो यस्य यच्च स्वलक्षणम् ।।२९।।
बीजानां प्ररोहाणाञ्च द्विगणे यच्च स्वरुपम् ।
अस्मिन् काण्डे बीजोत्पत्तौ तत् सर्वं सम्प्रकाशितम्।।३०।।

इति पराशरकृते वृक्षायुर्वेदे बीजोत्पत्तिकाण्डे द्विगणीय नाम अष्टमोऽध्यायः ।।३९।। इति बीजात्पत्ति काण्डः ।।

Bījotpatti Kāṇḍa
Chapter - VIII
Dwigaṇīya Adhyāya

Parāśara said : In this chapter of 'Dwigaṇīyadhyāya' I will delineate the subject on seeds and seedlings. A seed is the cause of perpetuating the continuity of plant kingdom. The inherent nature of a seed is to pierce upwards and sprout. [Verse 1-2].

Bījāngāni : Parts of a seed

Parts of a seed are termed as 'Kīkhosa', 'Bījamātṛkā', 'Bījapatra' and Mātṛkachhada 'Kīkhosa' is the outer cover of a seed (the seed coat).

The seed coat (Kīkhosa) according to its nature, is divided into three kinds, viz. 'Kāṣṭika', 'Karpara' and 'Mṛdula' meaning woody, brittle and soft respectively [Verse 3-4].

Bījamātṛkā :

'Bījamātṛkā' may be equated with the cotyledon of a seed, it is also termed 'Bījaśasyam' (Kernel).

Bījapatra : The first leaf inside the, 'Bījamātṛka' is called 'Bījapatra'. It may also be called Ādipatra (cf. Primary leaf) which has two other synonyms - 'Mātṛkacchada' and 'Kāñcuka' [Verse 5].

Seeds may be of two types - Ekamātṛk and Dwimātṛk Bīja. Ekamātṛkbīja has one cotyledon (monocotyledonous); and Dwimātṛkbīja has two cotyledons (dicotyledonous). These two types of seeds on germination show one or two leaves respectively [Verse 6].

Varṇata Bījāni : (Colour of seeds)

Fruits and seeds may have different colours due to the relative variation of the five elementary constituents of Pañcamahābhūta, viz. white, red, yellow, blue or variegated. [Verse 7].

Ākṛtito Bījāni : (Shape of seeds)

Seeds have different shapes. Some are long, some round, some roundish; others may be flat, some may be winged (Punkha Pucchika). Some may be coarse and others fine (Sthūla Sukṣmādīni). Again in some cases seeds are provided with sharp edged outgrowth. This kind of seed is called 'Dhārā-Bīja'. [Verse 8]. Seeds germinate on the soil in due course of time. In case of an epiphytic plant (Vṛkṣāruha) seeds germinate on the trunk of a tree ('Vṛkṣakāṇḍe prarohati'). [Verse 9].

Some plants ('Gulmakṣupaṇāṁ') have no flowering or fruiting property. Naturally, they are seedless. These plants are called 'Abakeśī'. The dorsal surface of the leaves of some 'Abakeśī' plants bear 'Patrānkuras' (cf. Spores) which have the property of seed ('Bījaswabhāvah'). By the help of these 'Patrānkuras', plants of the 'Abakeśī family propagate. [Verse 10-11].

There are some trees, creepers and shrubs which reproduce by branches or stem cuttings. They develop the 'Kandānkuras' or the vegetative buds on the stems. Some plants propagate by the vegetative buds borne on the stem (Kāṇḍānkuras) or on the rhizomes (Kandānkuras) or sometimes even on the roots (Mūlānkuras). They, naturally, function like seeds and thus have been discussed in this chapter on seed. [Verse 12].

Ovaries of some plants, despite having fertile flowers, do not produce complete (Pūrṇāgaṁ) seeds. The cause of such abnormal property is due to the diversion of the nourishing fluid ('Bījapauṣṭṛrasādayaśca') to the other parts of the plant body. Consequently, seeds in the ovary either do not form or remain in a rudimentary stage. And due to receiving in plenty of such diverted nourishing fluid, the other parts acquire the germinating property. This may be a rhizome (*Zingiber officinale*) or a tuber (*Solanum tuberosum*), a stem and even a leaf (*Saxifraga ligulata*). This kind of reproduction is respectively termed as 'Kandāruha', 'Kāṇḍāruha' and 'Patrāruha'. Some may call it an abnormal property. But this is not correct, as in such case the question would arise as to from where the first seed came ('Prathamabījābhāva'). [Verse 13].

Thus ends here the Bījaganīyasūtra.

Praroha Gaṇīya : The chapter on sprouting

Parāśara said : I will now discourse on sprouting. 'Praroha'. 'Udbheda' and 'Ankura' are synonymous and they all mean a sprout ('Praroha'). [Verse 14]. Different parts of a sprout are 'Bījamātṛkā', 'Bījapatra' and 'Mūlaṁ'. (They are equivalent to the cotyledon, the primary leaf and the root respectively). [Verse 15]. When a seed comes in contact with soil, it softens and swells. Next it manifests itself by striking a root into the soil. And thus it completes the sprouting process along with a visible cotyledon (or cotyledons) and sometimes with a primary leaf. [Verse 16].

Once the sprouting takes place, the seedling draws its nourishment (Rasa) from the Bījamātṛkā. With this supply of nourishment the seedling grows on. However, so long the root of the sprout has not fully established itself into the soil the process continues. [Verse 17]. By the time the established sprout is able to draw its nourishment directly from the soil ('Pārthiva rasa'), the cotyledons dry up (Bījamātṛkā praśoṣamāpadyate). Now the root draws the aqueous fluid from the soil ('Pārthiva rasa') and supplies it right upto the leaf through the 'Syandanī' (cf. vascular system). And here in the leaf 'Pārthiva rasa' gets digested in presence of the colouring matter ('Rañjakena pacyamānāt). In this process some sort of warm waste material (possibly gaseous) comes out (Yanmalaṁ nirgacchati taduṣmā bhavati).

The sap which is now a finer product nourishes the entire parts of the seedling . Consequently a seedling becomes a full fledged plant in due course ('Kāla pariṇāmāt sarvāngeṇa paripūrṇaṁ bhavati'). [Verse 18].

Shoots have been divided into two classes depending upon if they come out from a seed with one cotyledon or with two cotyledons, and they are named - 'Ekamātṛk' and 'Dwimātṛk' respectively. In case of an 'Ekamātṛk' shoot only one 'Ādipatra or primary leaf is found in the germinating seed, whereas in a 'Dwimātṛk' shoot there appears two 'Ādipatras' i.e., two primary leaves. Again by nature, a shoot or sprout may be of three kinds viz. 'Avyakta mātṛk', 'Vyakta mātṛk' and 'Vidala mātṛk'. An 'Avyakta mātṛk shoot' is that one whose cotyledon has a hard seed coat ('Kaṭhora Kīkhosena Mātṛkā saṁbita bhavati'). In case of a 'Vyakta Mātṛk' shoot, the seed coat being soft ('Mṛdula Kīkhosa'), the cotyledon becomes visible. In case of a 'Vidala Mātṛk' shoot, the cotyledon initially remains shrivelled ('Bīja mātṛka kuñcitam bhavati). After the shoot is fully nourished, the cotyledon becomes well formed. [Verse 19].

Shoots are of five kinds.

They are called : 'Bījānkur', 'Patrānkur', 'Kāṇḍānkur', 'Kandānkur', and 'Mūlānkur'. A shoot which germinates from a seed is called 'Bījānkur'. Likewise, a 'Patrānkur' means a shoot that sprouts from a leaf; Kāṇḍānkur means a shoot that is propagated by stem cutting; 'Kandānkur' refers to a shoot that sprouts out of a tuber or rhizome. And a 'Mūlānkur' means a shoot that comes out of a root. [Verse 20].

A plant may be called 'Bīja ruha', 'Patra ruha' 'Kāṇḍa ruha', 'Kanda ruha' and 'Mūla ruha' respectively depending on if the plant part is a seed (Bīja), a leaf (Patra), a stem (Kāṇḍa), a tuber or rhizome (Kanda) and a root (Mūla) as the organ of reproduction. [Verse 21].

During the process of germination of a seed there are four factors involved - soil, water, air and heat which help nourish the seedling and promote its growth. [Verse 22].

The soil provides anchorage ('Pṛthivī dhāraṇe') to the seedling; water is required for the synthesis of the materials from the soil ('Āpah sanśleṣaṇe'); air for the sprouting of the shoots ('Vāta udbhedane'); heat is necessary for the total thermal and colouring phenomena involved in the process of sprouting of a seed and thereafter for the growth of the plant. [Verse 23].

Parāśara said : I will now explain the different parts of a Praroha with respect to 'Pañcamahābhūta. 'Bīja mātṛkā' (Endosperm) and 'Mūla (Root) and odours of a sprout have earthy smell. Naturally they are concerned with the elementary 'Earth Mahābhūta' ('Pṛthivyātmakāni'). Likewise, the transport system ('Sanydanī') that carries aqueous fluid ('Rasa') and oleaginous matter ('Snehaśceti') is concerned with the 'Ap Mahābhūta' ('Abātmakāni'). Heat and colour which are associated with the leaves, originate from 'Teja Mahābhūta' ('Taijasatmakāni'). Similarly in the making of fibro-vascular bundle ('Sirā jāla') Vāyu Mahābhūta is concerned ('Vāyvātmakāni').

In the creation of an opening or hollowness (inside the sprout or other plant parts), the involvement of 'Ākāś Mahābhūta' (cf. Ethereal element) is imperative. [Verse 24].

As a result of any abnormal seasonal variation or variation in the proportion of Pāñcabhautik elements - either in lesser or greater degree or even if present in an unbalanced manner, - seeds may be defective in their different parts. Accordingly sprouts will show defects in their respective parts. [Verse 25].

Summary

Out of a jelly like substance, i.e., a colloidal agglomeration (which Parāśara defines as 'Piṇḍaṁ or 'Piṇḍasthānukaṁ' in the very beginning of the treatise — "Āpo hi kalalaṁ bhūtwa yat piṇḍasthānukaṁ bhavet"Bījotpattikāṇḍa, Ch. 1, Verse 6) the seed has its origin. A seed germinates into a sprout striking a root into the soil. The root bears leaves from which the stem develops; branches arise from the stem and bear flowers; flowers develop into fruits producing seeds. [Verse 26-27]. Thus the subject matter of 'Dwigaṇīyādhayāya' states as to how the germination of a seed maintains the natural process of continuation of plant life. [Verse 28].

Parāśara now summarizes the content of the Bījotpatti Kāṇḍa which are as follows: The origin of life; the soil character; distribution of forest regions in ancient India; plant organs - different types of leaves; various kinds of flowers and fruits; 'Aṣṭānga gaṇa saṁgraha', viz. root, bark, stem, heart-wood, sap, exudate (Niryās) fats and oils ('Sneha') and spine and prickle with their characteristics; and 'Dwigaṇīyādhyāya' with an account of different types of seeds and their nature of germination. [Verse 29-30].

Here ends the eighth chapter, the 'Dwigaṇīyādhyāya' of the Bījotpatti kāṇḍa of the Vṛkṣāyurveda by Parāśara.

वृक्षायुर्वेदः

वनस्पतिकाण्डः

वनस्पतिनिर्वर्णनाध्यायः

अथातो वनस्पतिनिर्वर्णनाध्यायं व्याख्यास्यामः इव्याह पराशरः । इह खलु वनस्पतिनाञ्चादौ सामान्यलक्षणमुपदिश्य विशेषेणोपदेक्ष्यते । वनस्पतिनान्तु पुष्पमव्यक्तं भवति । एव अपुष्पवन्तपुष्पेण संज्ञायते । यस्मात् वनस्पतिनां पुष्पाधारो वृत्तपरिमण्डलफल्गुस्थालकञ्च भवति। यतस्तत्र सूर्यो न भाति न चन्द्रमाश्च । सर्वे वनस्पतयः क्षीरवृक्षाश्चैव शुङ्गवृक्षा भवन्ति ।

तत्र श्लोकाः ।

अथ वनस्पतिश्चैव गुढपुष्पः क्षीरिद्रुमः ।
अपुष्पवन्तपुष्पेण यूस्तु हि परिकीर्तितः ।।
बीजमावर्तते यस्य स्थालके फल्गुसंज्ञके ।
तस्मात् फल्गुफलेनैव फल्गुद्रुम उदाहृतः ।।
यज्ञाङ्गः शुङ्गवृक्षश्च बहुबीजो द्विमातृकः ।
(Lacuna exists)..
सामान्यं नाम रुपञ्च फल्गुद्रुमस्य भाषितम् ।।
तत्र फल्गुद्रुमो यस्तु प्रकृत्या विविधःशृणु ।
कश्चिदवरोहिश्चैव कश्चिदेव विशिष्यते ।।
कश्चिच्छुक्लक्षीरिश्चैव कश्चिद् हेमक्षीरिर्भवेत ।
कस्यचित् कुक्षिफलञ्च काण्डे क्वचित् फलागमः ।
एवंविधश्च भेदश्च दृश्यते फल्गुपादपे ।।
अतो नामरुपाभ्याञ्च पृथक्त्वेन विवक्ष्यते ।

यस्य यद् नामरुपञ्च विशेषो यत्र यद् भवेत् ।।
वटः फल्गुस्थालकश्च रक्तफलः क्षीरिद्रुमः ।
कुक्षिफलः शुङ्गवृक्षः शुक्लक्षीरिर्महाद्रुमः ।।
न्यग्रोधपादपो ध्रुवः स्थिरायुर्जुहूपर्णकः ।
अवरोहिश्च स्कन्धजो बहुपादो वनस्पतिः ।।
अवटो नन्दीवटश्च नदीवटः क्षीरिद्रुमः ।
फल्गुफलो भवेच्चैव पीतफलो महाद्रुमः ।।
कुक्षिफलश्च शुङ्गी च शिल्पक्षीरिश्च स्कन्धजः ।
अवरोहिः प्ररोही च स्रुवपर्णो वनस्पतिः ।
पारिशः फलिशवृक्षः कुपीतनः खरच्छदः ।
फल्गुद्रुः पीतफलश्च कमण्डलुः सुदर्शनः ।।
अवरोहिः प्ररोही च क्षीरद्रुः कुक्षिफलदः ।।
इत्यवरोहिफल्गुद्रुमः। ।
अश्वत्थः पिप्पलश्चैव श्यामलः कुञ्जराशनः ।
शुभदः कुक्षिफलश्च फल्गुफलश्च याज्ञिकः ।।
श्रीपर्णो दीर्घवृन्तश्च चलपत्रः क्षीरिद्रुमः ।
वासन्तो मधुपर्णश्च पूतद्रु रक्तपल्लवः ।।
वटी पर्कटी प्लक्षश्च पिप्परिश्चारुदर्शनः ।
श्रीपर्णः क्षीरिवृक्षश्च फल्गुद्रुः कुक्षिफलदः ।।
इतिश्रीपर्णफल्गुद्रुमः ।
उदुम्बरः क्षीरिवृक्षो हेमदुग्धः सदाफलः ।
यज्ञाङ्गः काण्डफलश्च फल्गुद्रुर्यज्ञोदुम्बरः
इति हेमदुग्धकाण्डफलः फल्गुद्रुमः ।
नद्योदुम्बरः फल्गुश्च काण्डफलश्च शुङ्गकः ।
काकोदुम्बरः फल्गुश्च मलायुः कर्कशच्छदः ।
क्षीरद्रुः काण्डफलश्च फल्गुशाखः सदाफलः ।।
गुल्मोदुम्बरः फल्गुश्च कुक्षिफलः खरच्छदः ।
वल्लयोदुम्बरः फल्गुश्च कुक्षिफलश्च वारिजः ।

वल्लीफल्गुर्वलयश्च भूमिफल्गुः खरच्छदः ।।
अनेनोपदेशेन चानुक्तफल्गुफलानां गणविभागश्चादिशेत् ।
कस्मिंश्चित् फल्गुद्रुमगणे मिश्रलक्षणं भवति । तत्र फल्गुस्थालकेन सह पिण्डस्थालकमपि दृश्यते । एव फल्गुस्थालके परिमण्डलमध्यस्थे पिण्डस्थालकान्तिकैश्च कोषफलानि भवान्ति। एष गणो मिश्रफल्गुद्रुमो विज्ञायते ।
पनसः शुङ्गवृक्षश्च क्षीरद्रुः काण्डफलदः ।
मिश्रफल्गुगणश्चैव कोषफलो भवेच्च सः ।।
पिण्डफल्गुस्थालकाभ्यां कोषफलसमन्वितैः ।
द्विमातृकबीजश्चैव विषममातृकं भवेत् ।।
एवंविधं फलश्चैव मिश्रफल्गुगणः स्मृतः ।
इति प्रोक्तो वनस्पतिर्नामतश्च स्वरूपतः ।।

इति वृक्षायुर्वेदे वनस्पतिकाण्डे वनस्पतिनिर्वर्णाध्यायः ।।

VṚKṢĀYURVEDA
Vanaspati Kāṇḍa
Chapter - I
Vanaspatinirvarṇanādhyāya*

Parāśara said : I will now deal with the chapter on the plant division 'Vanaspati.'

In the beginning, I will give a general account of the plants of the 'Vanaspati'. Afterwards, their special characters will be delineated.

Flowers of Vanaspati plants are invisible ("Puṣpāmvyaktam bhavati"). They are also termed as 'Ăpuṣpavanta' - which means the flowers of Vanaspati plants do not get the exposure of lunar and solar rays, being enclosed within a 'Phalgusthālaka' the thalamus (Sthālaka) which is roundish and hollow ("Vṛttaparimaṇḍala phalgusthālakanca bhavati").

All the plants of Vanaspati order produce latex and are also characterized by a special type of leaf known as 'Śunga' ("Kṣīrivṛkṣaścaivaśungavṛkṣā bhavanti").

Summary

Now the characteristies of Vanaspati plants may be summarized as follows: They have flowers called 'Gudha puṣpa' (invisible flower) and 'Apuṣpavanta Puṣpa'. (Both

---

* Verse number was found missing in the manuscript.

the terms imply that the flowers do not get solar and lunar exposure). Secondly they have milky latex ('Kṣīridrumah'). Seeds of Vanaspati plants are borne in a 'Phalgu phala' (Hollow fruit). That is why a Vanaspati plant for its bearing a 'Phalgu phala' is called 'Phalgu drumah'. A Vanaspati plant or Phalgu druma bears many seeded fruits. Their seeds are dicotyledonous ("Bahu bīja dwimātṛkah"). The wood of some of the 'Phalgu druma' is used for kindling sacred fire of Vedic rites. That is why this plant is called 'Yājnānga' (e.g., the wood of *Ficus glomerata*). It bears the bud scales and is called Śunga vṛkṣa (Śunga-bud scale). Some of the Phalgu druma have adventitious roots, (e.g. *Ficus bengalensis*). Some of them have white latex. ('Śuklaksīriścaiva') and some others have yellowish latex ('Hemakṣīrirbhavet'). In some plants of the Vanaspati order fruits are borne at the leaf axils while in some cases they develop on the stem itself (as found in *Artocarpus heterophyllus*, and *Ficus glomerata*).

The above mentioned characteristics are found in the 'Phalgu pādapa.' I will now give examples of a few plants of the Vanaspati order with their names and the descriptive characters.

'Vaṭa' : (*Ficus bengalensis*, the Baniyan tree). Its fruit is red ('Rakta phalah'), with thalamus being round and hollow (Phalgusthālaka). The plant has got milky latex ('Ksīridrumah'). Fruits of Baniyan tree are borne in the leaf axil ('Kukṣiphalah'). The Baniyan tree has a special type of leaf (bad scale) which is called 'Śunga' ('Śunga vṛkṣa'). Its latex is white ('Śukla-Kṣīri'). The Baniyan tree has synonyms 'Mahādrumah' and 'Nyagrodha Pādapa'. It has got another synonym 'Sthirāyu' which means it has long life extending over a great many years. Its leaf is shaped like a 'Juhū' which is a special type of receptacle used in Vedic rites. (Note : Thus a 'Vaṭa' leaf is called 'Juhūparṇaka'). A Baniyan tree, has adventitious roots ('Avarohi') developed from the stems, that serve like so many supports for a large spreading tree ('Bahu pādapa Vanaspati'). Fig. 5a.

'Avaṭa' : It is another plant of Vanspati order. It has got several synonyms, viz., 'Nandī vaṭa', 'Nadī vaṭa', 'Kṣīridrumah', 'Phalguphalah', 'Pīta phalah', 'Mahā drumah', 'Kukṣiphalah', 'Śungī', 'Śilpakṣīrih', 'Skandhajah', 'Avarohih'. 'Prarohih' and 'Sruba-parṇah'.

'Pāriśa' : *(Ficus arnottiana)* 'Pāriśa' tree has numerous synonyms, viz. 'Phaliśa vṛkṣa', 'Kupītanah', 'Karacchadah', 'Phalgudruh', 'Pītaphalah', 'Kamaṇḍaluh', 'Sudarśanah', 'Avarohih', 'Prarohih', 'Kṣīradruh', 'Kukṣiphaladah'.

'Aśwattha' : (*Ficus religiosa*). It has got several synonyms viz., 'Pippala', 'Śyāmala', Kunjarāśana', 'Śubhadah', 'Kukṣiphalah', 'Phalguphalah', 'Yājnikah', 'Śrīparṇah', 'Dīrghavṛntah', 'Calapatrah', 'Kṣīradrumah', 'Vāsantah', 'Madhuparṇah', 'Pūtadruh' and 'Raktapallavah'. Fig. 5b.

Note : It is now quite apparent, that the above names, viz., Kukṣiphala, Phalguphala and Kṣīradruma are common synonyms, as found in the other plants of the Vanaspati order. The name 'Aśwatthah' is associated with religious sentiment held by the Hindus. Even in the Bhgvat Gītā, the Lord Śrī Kṛṣṇa says : 'Aśwatthah sarvavṛkṣāṇām' which means of all the trees Aśwattha has the superlative importance. (Bhgvat Gītā, Ch.X, śloka 26). Aśwattha tree has two more synonyms indicative of this sentiment. They are 'Śubhadah' and 'Pūtadruh'. Therefore, the latin equivalent *Ficus religiosa* is quite appropriate for this plant. The other synonyms are based on the general morphological characters. The synonym 'Śrīparṇa' satisfies the definition of it vide Vṛkṣānga Sūtrīyādhyāya, Ch. IV. Verse 52. The synonyms 'Dīrghavṛnta' and 'Calapatra' mean that the leaves of Aswattah tree have long petioles, and they are deciduous respectively. The other two synonyms, viz., 'Vāsanta' and 'Madhuparṇa' indicate that the new leaves appear in the spring time. The synonym 'Kuñjarāśana' means the leaves of Aswattha tree is one of the feeds of elephants. (Kuñjara means elephant).

'Plakṣa' (*Ficus talboti*) : This plant of Vanaspati order has few synomyms, viz., 'Pārkaṭī', 'Pippari', 'Cāudarśanah', 'Śrīparṇah;', 'Kṣīrivṛkṣah', 'Phalgudrumah', 'Kukṣiphaladah', the last four synonyms are common as found in the other plants of Vanaspati order.

'Udumvarah' (*Ficus recemosa*) : This plant has such synonyms as 'Kāṇḍaphala', Phalgudrumah'. 'Hemadugdah', 'Sadāphala', 'Yajnānga' and 'Yajnoudumvarah'. (Fig. 5c). The last two synonyms indicate that the wood of this plant is used for making sacred fire of Vedic rites. Further 'Udumvarah' has some other allied species, viz. 'Nadyodumvarah', Kākodumvarah' (*Ficus hispida*), 'Gulmaudumvarah', 'Vallyodumvarah'. These allied species have several synonyms, viz. 'Phalgu', 'Kāṇḍaphala', 'Kharacchadah' or 'Karkaśācchadah'.

Parāsara comments : there may be many more Vanaspati plants having mixed characters ("Miśra lakṣmaṇam bhavati") of Phalgusthālaka and Piṇḍa sthālaka. The latter remains enclosed within hollow cavity of the former and bears fruits termed Koṣa phala. These plants are grouped as Miśraphalgudrumah.

'Panasa' (*Artocarpus heterophyllus*) : It belongs to the group 'Miśra Phalgu gaṇa'. 'Panasa' has got several synonyms, viz. 'Śunga vṛkṣa', 'Kṣīradruh', 'Kāṇḍa phaladah' and 'Koṣa phala'. The 'Koṣa phala'[1] of Panasa tree is attached to the

---

1. Vide Ch. VI, verse 43, Bījotpaltikāṇḍa.
2. Vide Ch. V, verse 41, Bījotpaltikāṇḍa.
3. Vide Ch. V, verse 41, Bījotpaltikāṇḍa.

'Piṇḍasthālaka'[2] which is enclosed within 'Phalgusthālaka'.[3] It has got dicotyledonous seeds and cotyledons are of unequal size ("Viṣamamātṛkam bhavet"). Thus 'Panasa' is a plant typical of 'Miśra phalgu druma'. Fig. 5d

Here ends the Vanaspatinirvarṇanādhyāya of the Vanaspati Kāṇḍa of the Vṛkṣāyurveda by Parāśara.

वनस्पतिकाण्डः ।

वृक्षगणसंग्रहाध्यायः

अथातो वृक्षगणसंग्रहाध्यायं व्याख्यास्याम इत्याह पराशरः ।।१।। इह खलु चतुर्वर्गाणां वृक्षाणां वनस्पतिमधिकृत्य संग्रहेण गणविभाग उपदेक्ष्यते ।।२।। खलु द्वौ गणौ भवतः पर्णभेदेन जालिकपर्णगणीयश्च मौञ्जपर्णगणीयश्च ।।३।। द्वौ गणौ भवतः क्षीरेण ह्यक्षीरेण ।।४।। द्वौ गणौ भवतो व्यक्ताव्यक्तपुष्पाभ्याम् ।।५।। द्वौ गणौ भवतः समपुष्पविषमपुष्पाभ्याम् ।।६।। द्वौ गणौ भवतो व्यक्ताव्यक्तकेशराभ्याम् ।।७।। द्वौ गणौ भवतः सफलनिष्फलाभ्याम् ।।८।। द्वौ गणौ भवतः फलभेदेन कुड्यफलगणीयश्च विदरफलगणीयश्च ।।९।। द्वौ गणौ भवत एकबीजफलदश्च बहुबीजफलदश्च ।।१०।। द्वौ गणौ भवतो वर्तकभेदेन एकवर्तकगणीयश्च बहुवर्तकगणीयश्च ।।११।। द्वौ गणौ भवतो मातृकाभेदेन एकमातृकगणीयश्च द्विमातृकगणीयश्च ।।१२।। द्वौ गणौ भवतः प्ररोहभेदेन एकपर्णप्ररोहकश्च द्विपर्णप्ररोहकश्चेति संग्रहेण गणविभाग उक्तः ।।१३।।

भूयश्चात्र एतान् गणान् दृष्टान्तेन यथावदनुव्याख्यास्यामः ।।१४।। वनस्पतयो जालिकपर्णा भवन्ति । मौञ्जपर्णगणीयास्तु तालनारिकेलखर्जुरक्रमुक-तृणवल्वजादयश्च ।।१५।। वनस्पतयः क्षीरगणीयाश्चा अक्षीरगणीयास्तु अक्षपुष्पादयश्च ।।१६।। वनस्पतिनामव्यक्त पुष्पाणि भवन्ति । तेषां पुष्पं फल्गुस्थालकेन संवृतं भवति । वानस्पत्यानान्तु व्यक्तपुष्पाणी स्युः ।।१७।। चूतपुष्पगणीयश्च समपुष्पं भवति । वनस्पतिनान्तु विषमपुष्पञ्च ।।१८।। समिपुष्पाणां व्यक्तकेशरश्च। मल्लिकापुष्पाणामव्यक्तकेशरं भवति ।।१९।।

वनस्पतयः सफलगणीयाश्च । मञ्जुपुष्पिणश्च निष्फला भवन्ति ।।२०।। वनस्पतिनां कुड्यफलं भवति । आस्फोटानां विदरफलञ्च ।।२१।। चूतपुष्पवृक्षाणामेकवीजफलं भवति । वनस्पतयो बहुबीजफला भवन्ति ।।२२।। वनस्पतय एकवर्तकफला भवन्ति । ओड्रपुष्पगणीयानां विदरफलञ्चैव बहुवर्तकफलञ्च ।।२३।। एकमातृकगणीयश्च तृणवल्वजादयश्चा द्विमातृकगणीयास्तु वनस्पतयः ।।२४।। तृणवल्वजकण्डुलानां प्ररोहे ह्येकपत्रं भवति । वनस्पतिनां द्विपत्रञ्च ।।२५।। इति पराशरकृते वृक्षायुर्वेदे वनस्पतिकाण्डे वृक्षगणसंग्रहाध्यायोनाम द्वितीयोऽध्यायः ।

Vanaspati Kāṇḍa
Chapter - II
Vṛkṣagaṇasamgrahādhyāya

Parāśara said - I will now explain Vṛkṣagaṇasamgrah. [Verse 1].

I will deal with the classification of four groups of plants with special reference to the group 'Vanaspati'.

(Most plants may be grouped on the basis of pairs of opposite or contrasting characters, which are as follows)

(a) Plants may have leaves with reticulate venation (Jālika parṇa gaṇīya) or parallel venation (Mauñja parṇa gaṇīya). [Verse 3].

(b) Plants having latex or not ("Dwaugaṇaubhavatah Kṣīreṇhyakṣīreṇa"). [Verse 4].

(c) Whether plants have floral members properly manifest or not. [Verse 5].

(d) Plants may have flowers isomerous or heteromerous. [Verse 6].

(e) Whether stamens are protruding or enclosed within the corolla. [Verse 7].

(f) Plants having fertile or sterile flowers. [Verse 8].

(g) Plant may have fruit dehiscent or indehiscent.
[Verse 9].

(h) Plants may have fruit with a single seed or many seeds. [Verse 10].

(i) Whether the ovary is unilocular or multilocular. [Verse 11].

(j) Whether the seed is monocotyledonous or dicotyledonous. [Verse 12].

(k) Seedling with a single seed leaf or two seed leaves. [Verse 13-14].

I will now explain this classification through examples. Vanaspati plants have reticulate leaves, but plants like - Palymra (*Borassus flabellifer*), Coconut (*Cocos nucifera*), Date palm (*Phoenix dactylifera*), Betel nut (*Areca catechu*) etc. and other plants of the grass family and the Palmae family ('Tṛṇavalwajadayasca') have parellel venations. [Verse 15]. Vanaspati plants have latex, and plants of the Akṣapuṣpagaṇa (Combretaceae family) have no latex. [Verse 16]. Vanaspati plants have flowers which do not get solar or lunar exposure. Flowers of this category remain hidden inside the 'Phalgusthālaka', whereas flowers of Vānaspatya group are uncovered ('Vyaktapuṣpāṇisyuh). [Verse 17]. Flowers of the Cūtapuṣpa gaṇa (Anacardicaeae family) are isomerous while plants of Vanaspati group have heteromerous flowers. [Verse 18]. Flowers of the Samipuṣpa gaṇa (Leguminosae family) have their stamens ('Vyaktakeśaraśca') open or protruding out with the opening of the petals, while in case of the flowers of the Mallikāpuṣpa gaṇa (Apocynaceae family), stamens remain hidden inside the corolla ('Avyaktakeśara'). [Verse 19]. Flowers of Vanaspati plants are fertile, while Mañju puṣpa are sterile and do not produce any fruit [Verse 20]. Fruits of Vanaspati plants are indehiscent. Fruits of the Cūtapuṣpa gaṇa (Anacardiaceae) have only one seed, whereas fruits of Vanaspati plants are many seeded. [Verse 22]. Fruits of Vanaspati plants are uniocular. In case of the Oḍrapuṣpa gaṇa (Malvaceae) fruits are dehiscent and the ovary is multilocular. [Verse 23].

In plants of the grass and the Palmae family seeds are monocotyledonous, whereas seeds of Vanaspati plants are dicotyledonous. [Verse 24].

On sprouting, the plants of the Araceae family ('Tṛṇavalwajakaṇḍulānam') show only one leaf, whereas Vanaspati plants show two leaves. [Verse 25].

Here ends the second chapter, the 'Vṛkṣagaṇasamgrahādhyāya' of the Vanaspati Kāṇḍa of the Vṛkṣāyurveda by Parāśara. [Verse 26].

वनस्पतिकाण्ड :

तृणवर्गाध्याय:

अथातस्तृणराजवर्गाध्यायं व्याख्यास्याम इत्याह पराशरः । खलु तृणवृक्षास्तु मौञ्जपर्णमेकमातृकमंशुकं फलवल्कलञ्च भवन्ति । ते एव त्वक्सारश्च । तृणवृक्षास्तु त्रिविधा भवन्ति यथा कुक्षिमञ्जरिकश्च शीर्षमञ्जरिकश्च कन्दवल्वजाश्चेति । तत्र तृणराजानस्तालनारिकेलखर्जुरक्रमुकवेत्रप्रभृतयो बहुवर्षजीविनश्च । एतेषां पुष्पमञ्जरिः पत्रपट्टिककुक्षौ काण्डगात्रे सञ्जायते । एव बीजमातृकाया आवरणं कठोरं कर्परं भवति । कुक्षिमञ्जरिकश्च पुनर्द्विविधो भवति यथैकबीजफलदो बहुबीजफलदश्चेति । तत्र नारिकेलखर्जुरक्रमुकादयश्चैकबीजफलदा भवन्ति । तालस्तु बहुवीजफलदश्च। एव सफलं निष्फलञ्च । निष्फले मार्जारपुच्छिकं मञ्जरिर्भवति । शीर्षमञ्जरिकतृणानान्तु काण्डो बहुसन्धितो भवति । तस्मात् पर्विकं भवत्येव निःसारञ्च। एतेषां काण्डः पत्रपट्टिकेन संवीतो भवति । एवञ्च शीर्षग्रन्थौ पुष्पमञ्जरिः सञ्जायते । तस्मात् शीर्षमञ्जरिक आचक्षते, एवोषधिर्विज्ञायते फलपाकान्ताच्च । एतेषां केचिद् बीजरुहाः केचित् मूलपर्वरुहाः केचित् काण्डरुहाश्च ।

तत्र धान्ययवगोधुमश्यामाककोद्रववज्जिरादयश्च बीजरुहा भवन्ति । वंशनलमहानलपोटगलादयश्च मूलपर्वरुहाश्च भवन्ति । एवंविधा ये तृणास्तेऽपि । इक्षुकाण्डिरिक्षुवाटिकाकाशादयश्च काण्डरुहा भवन्ति च्छिन्नरुहा एव विज्ञायते ।

कन्दवल्वजस्तु तृणलक्षणेन कान्दिकमूलं भवति । एतेषां पुष्पप्रभवं काण्डञ्च भवति यच्चाग्रमञ्जरिको विज्ञायते।

कन्दवल्वजस्तु द्विविधो भवति यथा कन्दप्रोहिकश्च कुक्ष्याङ्कुरिकश्चेति । तत्र कन्दप्रोहिकास्तुकन्दगात्रात्समुद्भवन्ति। यथा कदलीशृङ्गबेरशठि-हरिद्राकर्चुरवचादयश्च । एष गणो मोचिकपुष्पमाचक्षते । एतेषां मोचिकेन पुष्पं दृश्यते । कुक्ष्याङ्कुरिकस्तु पट्टिककन्दकुक्षौ प्ररोहः सञ्जायते । यथा रसोनपलाण्डु नक्तगन्धादयश्च । एवंविधाश्च ये कन्दास्तेऽपि कुक्ष्याङ्कुरिककन्दवल्वजाश्च । एष गणाः पीवरपुष्पमाचक्षते । अनेन विधिनातृणवर्गो व्यावस्येत् ।

इति पराशरकृते वृक्षायुर्वेदे वनस्पतिकाण्डे तृणराजवर्गाध्यायो नाम तृतीयोऽध्यायः ।

Vanaspati Kāṇḍa
Chapter - III
Tṛṇavargādhyāya*

Parāśara said : I will now delineate the plants of the grass and the Palmae family ('Tṛṇarājavargādhyāya'). Plants of these families have leaves with parallel venation ('Mauñjaparṇa'), seeds are monocotyledonous and their fruit walls are fibrous ('Angśukaphalavalkal'). Some plants (like *Bambusa arundinacea*), have their bark which is substantially thick. The plants of the grass and the Palmae family have three kinds of inflorescences, viz. i) an inflorescence that comes out of the leaf-axil ('Kukṣimañjari') and ii) the inflorescence is terminal ('Śīrṣamañjari') and iii) borne on the rhizome.

Note : As example of 'Tṛṇarājvṛkṣa' Parāśara cites - *Borassus flabellifer, Cocos nucifera, Phoenix dactylifera, Areca catechu* and *Calamus rotang.* These trees live for many years.

The inflorescences in some of these plants are borne at the leaf axils of the stem.

In some plants of the grass and the Palmae family the seed coat is hard. Inflorescences, coming out of the leaf axils, are of two typeṣ. Some have fruit with a single seed and

---

* Verse number was found missing.

some with many seeds. In case of Narikel (*Cocos nucifera*), Kharjur (*Phoenix dactylifera*), Kramuka (*Areca catechu*) the fruits are one seeded only, whereas Tāla (*Borassus fabellifer*) bears many seeded fruits; this plant has both fertile (hermaphrodite) and sterile flowers. The inflorescence of this plant bearing sterile flowers looks like the tail of a cat (cf. Catkin).

The stems of plants of the grass family, with terminal inflorescence, have numerous segments with joints '(Parva') and these are hollow (e.g. *Bambusa arundinacea*). Their stem is ensheathed by a spathe. The inflorescence is borne terminally on the stem and is thus called 'Sīrṣamañjarik'. The plants die when their fruits/seeds ripen. Of the plants belonging to the grass family, some are propagated by seeds, some by roots and some by segemnts of stems. Plants which are propagated by seeds are 'Dhānya' (*Oryza sativa*). 'Yava' (*Hordeum* vulgare), 'Godhuma' (*Triticum aestivum*), 'Śyamā' (*Echinochloa frumentacea*). 'Kakodrava' (*Paspalum scrobiculatum*), and 'Bajjira' (*Pennisetum amerecanum*) 'Ḅanśa' '(*Bambusa* sp.), 'Nala' (*Phragmites maxima*). 'Kāś' (*Saccharum spontaneum*) are propagated by planting their roots and segments, whereas 'Ikṣu' (*Saccharum arundinaceum*) is propagated by stem. They are also termed 'chhinnaruha'.

Like grasses plants of the Araceae family have rhizomes or bulbs (Kandavalwaja). Their inflorescences are protected by the Spadix (Akṣamañjari, vide, Puṣpāngasūtrīyādhyāya, Ch. V, Verse 28).

Plants having rhizomes and bulbs are of two types viz., 'Kanda prarohikā' and 'Kukṣyānkurikā'. The former develops from the rhizome body ('Kandagātrāt'). As for examples, 'Kadali' (*Musa* sp.), 'Sṛngaber' (*Zingiber officinale*) 'Saṭi' (*Curcuma zedoaria*), 'Haridrā' (*Curcuma longa*), 'Karcur' (*Hedhychium spicatum*), 'Vaca' (*Acorus calamus*). Inflorescences of these plants have spathe; and plants like 'Rason' (*Allium sativa*), 'Palāṇḍu' (*Allium cepa*),

'Naktagandhā' (*Polianthes tuberosa*), grow out of 'Kukṣyānkurikā' that is spadix inflorescence. Flowers of these plants are also called 'Pīvarapuṣpa'. Thus based upon the above characteristics plants of the grass family are identified.

Here ends the Tṛṇavargādhyāya, the thind chapter of the Vanaspati Kāṇḍa' of the Vṛkṣāyurveda by Parāśara.

वानस्पत्यकाण्डः

वानस्पत्यविचरणाध्यायः ।

अथातो वानस्पत्यविचरणाध्यायं व्याख्यास्याम
इत्याह पराशरः ।।१।।
व्यक्तपुष्पैः फलैर्येऽपि दृश्यन्ते वनपादपाः ।
संज्ञितास्ते वानस्पत्यास्तेषां वर्गो विवक्ष्यते ।।२।।
आम्राम्रातकजिङ्गनभल्लातकश्च पादपाः ।
एष गणश्चूतपुष्पं भवति तुन्दमण्डले ।।३।।
यस्य पञ्चदलं पुष्पं पञ्चैव केशरान्वितम् ।
क्षुल्लकं पुष्पमण्डलं वर्तकमेकबीजकम् ।
कुड्यफलं भवेत्तत्र बीजं वृन्तान्तिकं तथा ।।४।।
आम्रश्चुतो रसालश्च पिकप्रियो मधुद्रुमः ।
जतुद्रुः सहकारश्च सौरभो भृङ्गवल्लभः ।।५।।
आम्रातको भृङ्गीवृक्षः कपिचूतः पीतनकः ।
छिन्नरुहो भवेद् वृक्षश्चाम्लवटो मधुच्छदः ।।६।।
जिङ्गनश्छिन्नरुहश्च साटकश्च सुनिर्यासः ।।७।।
भल्लातको जतुवृक्षो वह्निमुखश्चारुष्करः ।
एतैश्च नामलिङ्गैश्च कीर्तिताश्चुतवर्गीयाः ।।८।।
इत्याम्रादिवर्गः ।

कीलकगन्ध्पर्णानां वृक्षाणामुपपक्षाणाम् ।
दृश्यते पुप्लिकपुष्पं तेषां नाम विवक्ष्यते ।।९।।

द्विविधो भवति भेदः फलभेदेन पुप्लिकम् ।
कस्यचित् फलशस्यन्तु दृश्यते केशरनिभम् ।
कस्यचित् फलशस्यञ्च संहतञ्च घनं भवेत ।।१०।।
आद्यं केशरकं ज्ञेयं द्वितीयं मालुरफलम् ।।११।।
मातुलुङ्गनागरङ्गजम्बिरनिम्बुकादयः ।
भवन्ति पुप्लिकपुष्पं केशरकं फलञ्च तत् ।।१२।।
मालुरं भवति विल्वः कपित्थश्च तथैव च ।।१३।।
मातुलुङ्गं बीजपूरं पृथुकश्चास्य बीजकम् ।।१४।।
नागरङ्गं मधुरञ्च सुरङ्गं स्वर्णवल्कलम ।।१५।।
जम्बिरं दन्तहर्षकश्चात्यम्लं मुखशोधकम् ।।१६।।
निम्बुकं क्षुद्रजम्बिरं सुगन्धं मुखरोचकम् ।।१७।।
विल्वञ्च श्रीफलं प्रोक्तं मालुरञ्च महाफलम् ।
त्रिपर्ण गन्धपर्णञ्च गन्धफलश्च शल्यकम् ।।१८।।
कपित्थं दधिफलञ्च मालुरं सुरभिच्छदः ।
समाहृत इति प्रोक्तो मातुलुङ्गादिवर्गो वै ।।१९।।

इति मातुलुङ्गदिवर्गः ।

जम्बुकादीनि फलानि पारेवतफलञ्च यत् ।
दृश्यते देवपुष्पेण यच्चैव कुम्भमण्डलम् ।।२०।।
देवपुष्पं भवेत्तत्र मुक्तञ्च कुम्भमण्डलम् ।
अशाखमञ्जरिः काचित् काचिच्छाखासमन्विता ।।
जालकं युक्तकर्णं वा यच्चतुष्कमथापि वा ।
दलञ्चैवभवेद् मुक्तं तत्रैव बहुकेशराः ।।
कुड्याख्यञ्चैव फलञ्च वर्तकमेकमेव वा ।
क्वचिद् वृन्तातिकं बीजं क्वचिदक्षान्तिकं भवेत ।।
एकञ्च कस्यचिद् बीजं बहुबीजानि कस्यचित् ।।२१।।
जम्बुस्तु बहुधा ज्ञेयास्तेषां भेदमपि ब्रुवे ।
राजजम्बुः सुवर्णाभा फलेन्द्रा सुरभिफला ।।
वीरपत्रा बृहत्फला महाजम्बुः प्रकीर्तिता ।।२२।।

मेघाभा कृष्णजम्बुस्तु शुक्तिपर्णा तु मध्यमा ।
नादेयी वेतसपर्णा वनजात्यम्लजम्बुका ।।२३।।
मधुरा क्षुद्रजम्बुस्तु काकजम्बु: प्रकीर्तिता ।।२४।।
पारेवतं रैवतकं मधुराम्लफलञ्च तत् ।
सितारुणशलाटुभ्यां द्विविधं भिद्यते फलम् ।।
समाहृत इति वर्गो जम्बुकादिविनिश्चय: ।।२५।।
इति जम्बुकादि वर्ग: ।

हरितकीविभितकीवातामार्जुनपादपा: ।
विज्ञेया अक्षपुष्पेण ये चान्यतुल्यलक्षणा: ।।२६।।
हरितक्यभया प्रोक्ता कषाया हैमवत्यपि ।
अस्त्रकफलबीजञ्च दीर्घपर्णं मधुद्रुम: ।।२७।।
विभितक: कर्षफल: कषाय: कर्परबीजम् ।
वासन्तो मधुपर्णश्च स्निग्धमज्जा च तैलद: ।।२८।।
वातामं मधुपर्णञ्च नेत्रोपमफलञ्च वै ।
कर्परश्च भवेद् बीजं स्निग्धमज्जा हि तैलदम ।।२९।।
अर्जुन: ककुभ: प्रोक्तो वासन्तोऽरुणवल्कलम् ।
धवलो धाराफलश्च धुम्रसारश्च बीजक: ।।३०।।
कदाचिदक्षपुष्पञ्च वल्लर्यां निष्फलं भवेत् ।
सफलञ्च यदा पुष्पं तदेव कुम्भमण्डलम् ।।३१।।
इति हरितक्यादि वर्ग: ।

कोलश्च बदरश्चैवभवत: कण्टकद्रुमौ ।
एष वर्ग: कुहपुष्पं भवति तुङ्गमण्डलम् ।।३२।।
कोलं सौवीरकं प्रोक्तं मधुरं जाङ्गलोद्भवम् ।
बदरमम्लमधुरं फलञ्च पिच्छिलं भवेत् ।।३३।।
इति कोलादिवर्ग: ।

इति वृक्षायुर्वेदे वानस्पत्य काण्डे वानस्पत्यविचरणाध्यायो नाम प्रथमोऽध्याय: ।।३४।।

अथ सप्तपर्णादिवर्गः ।

सप्तपर्णः कुटजश्च पुष्पेण मल्लिका स्मृता ।
तत्राराबन्धनं विद्यात् सप्तपर्णे महाद्रुमे ।।
कुटजे पक्षपङ्क्तिक्रं पत्रवृन्तेन विज्ञेयम् ।।
शरदि सप्तपर्णश्च कुटजः प्रावृषायणे ।
पुष्पितो भवति वृक्षो वानस्पत्येन संज्ञितः ।।
अत ऊर्ध्वं संक्षेपेणं शृणु तत् पुष्पलक्षणम् ।
पुष्पाङ्गसूत्रीये यच्च विस्तारेण विवक्ष्यते ।।
मल्लिका कुण्डमण्डला संकुला पत्रवल्लरिः ।
अरापक्षपङ्क्तिका वा समाङ्गा पुष्पगणीया ।।
जालकं युक्तं कर्णञ्च पञ्च भवति तत्र वै ।
पञ्च युक्तं दलश्चैव केशरपञ्चभिः सह ।।
वराटद्वययुक्ता च युग्मफला भवेच्च सा ।
सक्षीरा बहुबीजा च तुलपुच्छसमन्विता ।।
सप्तपर्णो बृहद्द्रुमस्तिक्तदुग्धः सुदर्शनः ।
उद्यानभूषणश्चैव शारदो मदगन्धकः ।।
कुटजो वत्सकः शक्रः कालिङ्गो गिरिमल्लिका ।
प्रावृष्यो यवबीजश्च तिक्तद्रुः क्षीरपादपः ।
एतैश्च नामलिङ्गैश्च विज्ञेयो मुनिसत्तम ।।

समिवर्गाध्यायः ।

अथातः समिवर्गाध्यायं व्याख्यास्याम इत्याह पराशरः ।।१।।
समिवृक्षान् प्रवक्ष्यामि येषां सिम्बिफलं भवेत् ।
सिम्बिपर्णं क्वचित्तेषां पुङ्खपर्णं तथैव च ।।
माढिपर्ण भवेच्चैव रविकान्ता समिः स्मृता ।
इमे तु समिवृक्षाश्च प्रायशो जाङ्गलोद्भवाः ।।२।।
समीनां त्रिविधो भेदो यदुक्तं पुष्पसूत्रीये।
दृष्टान्तेन समुद्दिष्टमिह यद् यस्य लक्षणम् ।।३।।

वक्रपुष्पं विकर्णञ्च शूकपुष्पं च यद् भवेत् ।
तदेव वर्गभेदेन त्विहानु वक्ष्यते शृणु ।।४।।
अत ऊर्ध्वं प्रवक्ष्यामि वक्रपुष्पस्य लक्षणम् ।
वक्रपुष्पं भवेत्तत्र विषमतुन्दमण्डलम् ।।
जालकं युक्तकर्णञ्च पञ्च भवति तत्र वै ।
विषमदलसंस्थानं वक्रेण पञ्चकं भवेत् ।।
दलन्तु भवति मुक्तं केशरैर्दशभिस्तथा ।
तेषान्तु नवयुक्तं च भवति पुष्पमण्डले ।।
सिम्बिफलं भवेत्तत्र पार्श्वबीजं समन्वितम् ।।५।।
विकर्णपुष्पमण्डलं विषमतुन्दसंज्ञितम् ।
जालकं भवति युक्तं पञ्चकर्णं समन्वितम् ।।
दलन्तु भवति मुक्तं विषमकर्णसंस्थानैः ।
दशभिः केशरमुक्तैर्भवति पुष्पलक्षणम् ।।६।।
पञ्चभिः केशरैः क्वचिद् दृश्यते पुष्पमण्डलम् ।।७।।
शूकाकारं भवेत् पुष्पं तस्मात्तु शूकसंज्ञितम् ।।८।।
अत ऊर्ध्वं पृथक्त्वेन वक्ष्यन्ते समिजातयः ।
लक्षणेन प्राकृतेन नाम्ना हि संस्कृतेन च ।।९।।
पलाशः सिम्बिपर्णञ्च वक्रपुष्पञ्च किंशुकः ।
क्षारश्रेष्ठो जाङ्गलश्च याज्ञिको ब्रह्मपादपः ।।१०।।
अगस्त्यो पुङ्खपर्णञ्च वक्रपुष्पं मुनिद्रुमः ।।११।।
जयन्ती पुङ्खपर्णी च नादेयी वक्रपुष्पिका ।।१२।।
करञ्जो नक्तमालश्च वक्रपुष्पञ्च पुतिकः ।।१३।।
शिंशपा वक्रपुष्पी च कृष्णसारा महाद्रुमा ।।१४।।
पारिभद्रः सिम्बतरुः सिम्बिपर्णेन दृश्यते ।।१५।।
एते वै पादपाः सर्वे वक्रपुष्पेण दृश्यन्ते ।।१६।।
आरग्वधो राजवृक्षः पीतपुष्पविकर्णिकः ।
कपिदण्डफलश्चैव स्वर्णाटुश्चतुरङ्गुलः ।।
सम्पाकः कृतमालश्च पुङ्खपर्णं भवेद् द्रुमः ।।१७।।

कोविदारो युग्मपत्रं श्वेतपुष्पविकर्णकम् ।
पञ्चकर्णं भवेत पुष्पं पञ्चैव मुक्तकेशरम् ।।१८।।
काञ्चनारो युग्मपत्रं विकर्णताम्रकुसुमम् ।।१९।।
अश्मन्तको युग्मपत्रं पीतपुष्पविकर्णिकम् ।।२०।।
अंशुकः पूतवृक्षश्च तन्तुलं श्लक्ष्णवल्कलम् ।
एते वै पादपाः सर्वे विकर्णपुष्पसंज्ञकाः ।।२१।।
शिरीषः शूकपुष्पञ्च पुङ्खपर्ण शुकप्रियः ।
उद्यानको दृढसारः श्यामलः सिम्बिनीफलः ।।२२।।
खदिरः शूकपुष्पञ्च बालाह्वपुङ्खबन्धनम् ।
गायत्री रक्तसारश्च कण्टकी याज्ञिकः स्मृतः ।।२३।।
कदरः सोमवल्कश्च खदिरोऽन्यः पथिद्रुमः ।।२४।।
समी तु बालपत्री च पुङ्खपर्णी समिद्वरः ।
शूकपुष्पी भवेत्तुङ्गा पवित्रा शङ्कुकलिका ।।२५।।
अरिस्तु शूकपुष्पी च पीतपुष्पी सुगन्धिका ।
पुङ्खपर्णी भवेत् सा तु बालपत्री च मरुजा ।।२६।।
काम्भोजी विट्खदिरश्च दुर्गन्धा पीतपुष्पिका ।
खदिरादयः सर्वे तु कषायाः कण्टकद्रुमाः ।।२७।।
शिरीषादयः सर्वे तु शूकपुष्पेण वीक्षिताः ।।२८।।
अशोको वञ्जुलं ज्ञेयं फलेन सिम्बिका भवेत् ।
गुच्छमञ्जरिका तत्र वासन्तो रक्तपल्लवः ।।२९।।
श्योनाको दीर्घसिम्बिः स्याद् वञ्जुलो भूतपुष्पकम् ।
माढिना मञ्जुपर्णकं कट्वङ्गः पीतवल्कलम् ।।३०।।
तिन्तिरी मिश्रसिम्बिः स्यात् पुष्पेण तु विशिष्यते ।
अम्लिका चाम्लफलञ्च वृक्षाम्लं क्षारपादपा ।।३१।।

VṚKṢĀYURVEDA
Vānaṣpatyakāṇḍa
Chapter - I
Vānaspatya Vicaraṇādhyāya

Parāśara said : I will now deal with those plants which come under the group termed 'Vānaspatya'. [Verse 1].

Plants in which flowers are visible and which bear fruits are termed 'Vānaspatya' which will now be described. [Verse 2].

Cūtapuṣpagaṇa : Āmra (*Mangifera indica*) Āmrātaka (*Spondias pinnata*), Jingan (*Lannea grandis*), Bhallātaka (*Semecarpus ancasrdium*). etc. ("Āmrararātakajingana-bhallātakaśca") belong to the Cūtapuṣpa gaṇa (cf. Anacardiaceae) The flowers of this family are hypogynous (cf. Tundamaṇḍala, Bījotpattikāṇḍa, Ch. V, Verse 58). [Verse 3].

Flowers of Cūtapuṣpa gaṇa have 5 petals and 5 stamens; flowers are small and have unilocular ovary. Fruits are one seeded. Fruit is indehiscent and the seed is oriented towards the pedicel. [Verse 4].

'Āmra '(*Mangifera indica*). - The fruit is very juicy and tasty and is liked by cuckoos ('Pikapriya'). Flowers of *Mangifera indica* appears in spring ('Madhudrumah'). Mango tree produces a kind of oleroesinous exudate (Jatudru). The tree has a synonym 'Sahakāra'; flowers have some fragrance which attracts wasp like flying animates ("Saurabhobhṛnga Vallabha".). [Verse 5].

'Āmrātaka(*Spondias pinnata*) has three synonyms viz. 'Bhṛngī-vṛkṣah', 'Kapicutah' and 'Pītanakah'. (Note : Of these synonyms the first one, viz. 'Bhṛngīvṛkṣa' has been explained before. The second synonym viz. 'Kapicutah', means that the fruit of *Spondias pinnata* is liked by monkeys. The third synonym, viz. 'Pītanaka' could not be deciphered). The tree, *Spondias pinnata* can be propagated by cuttings (chhinnaruha); the fruit is sour and leaves appear in spring. [Verse 6].

The tree of Jingan (*Lannea grandis*) exudates a fragrant fluid which has got adhesive property ("Sāṭakaśca suniryāsah"). [Verse 7].

The plant 'Bhallātaka' (*Semecarpus anacardium*) has synonyms, viz. 'Jatuvṛkṣa', 'Vanhimukha' and 'Aruskar' (fruits having a substance with a corrosive and inflammable property). With the above descriptions and some names the 'Cūtapuṣpagaṇa' (cf. Anacārdiaceae) has been defined. [Verse 8].

Here ends the 'Āmrādivarga' of 'Cūtapuṣpagaṇa'.

Puplikpuṣpagaṇa : The plants of the Puplikpuṣpagaṇa (cf. Rutaceae family) have spines ('Kīlaka') and their leaves are aromatic and bear stipules—'Upapakṣam'-(Vide Bījotpattikāṇḍa, Ch. IV, Verse 32). [Verse 9].

Parāśara said : I will now narrate the plants of the Puplikgaṇa Fruits of the Puplikgaṇa are of two types. Of them one kind has the edible portion resembling stamens. The other kind is thick and pulpy. [Verse 10]. The former type is known as 'Keśarakaṁ' (vide Bījotpattikāṇḍa, Ch. VI, Verse 42, e.g. Citrus fruits); and the latter is known as

'Māluraphalam'. [Verse 11]. Various Citrus fruits, viz. 'Mātulunga', 'Nāgaranga', 'Jambira', 'Nimbuka', etc. belonging to the Puplikagaṇa (Rutaceae family) fall under the group of 'Keśarakaphal'. [Verse 12]. Examples of Māluraphalaṁ are : Bilwa (*Aegle marmelos*) and Kapittha (*Feronia elephantum*). Mātulunga (*Citrus medica*) has a synonym viz. Bījapur. Its seeds are flattened. [Verse 14].

The Fruits of 'Nāgaranga' (Citrus reticulata) is sweet in taste. It has synonyms 'Surangam' and 'Swarṇavalkalam' (the fruit wall is golden in colour). [Verse 15]. 'Jambira' (Citrus lemon) is very sour causing a tingling sensation in teeth. It also deodorizes the mouth. [Verse 16]. 'Nimbuka' (Citrus aurantifolia) is small, aromatic and appetising. [Verse 17]. The 'Bilwa' (*Aegle marmelos*) has the synonyms - 'Śrīphalaṁ' and 'Mahāphalaṁ'. The leaf of *Aegle marmelos* is trifoliate ('Triparṇam') and it has an aroma. Its fruit is also aromatic ("Gandhaphalanca"). The tree has got spines ('Śallyakaṁ') [Verse 18]. *Feronia elephantum* the 'Kapittha' synonymous with 'Dadhiphala' and 'Mālura' belongs to the Puplikgaṇa (Rutaceae family). Its leaf is aromatic ('Surabhicchadah').

Here ends the Mātulungādi varga of the family Puplikpuṣpagaṇa. [Verse 19].

Devapuṣpagaṇa

'Jambuka' (*Syzygium cumini*) and 'Pārevata' (*Psidium guajava*) belong to the 'Devapuṣpagaṇa' (cf. Myrtaceae family). [Verse 20].

Flowers of the Devapuṣpagaṇa are polypetalous and epigynous (Kumbhamaṇḍala, vide Bījotapattikāṇḍa Ch. V. Verse 58). In some cases inflorescences are unbranched ('Aśākhamañjari') and in others they are branched ('Saśākha'). The calyx is gamosepalus and appears husk like. The corolla is polypetalous; stamens are many. The fruit of the Devapuṣpagaṇa is indehiscent ('Kuḍyākhyañcaiva') and it has unilocular ovary ('Vartakamekameva'). In some cases fruits have seeds which are attached to the pedicel ("Vṛntāntika bījam") and in others they are close to the floral axis (Akṣāntik). Plant produces both one seeded and many seeded fruits. [Verse 21]. There are various kinds of *Syzygium* species. Of which, 'Rājajambu' has fruits of golden yellow colour ('Suvarṇābha) with aroma. It is also called 'Mahājambu' as it produces big fruits; its leaves are long ('Vīrapatrī'). [Verse 22]. 'Kṛṣṇa-jambu' has blackish fruits; its leaf has the outline of a mother of pearl (Śuktiparṇa). It produces medium sized fruits. 'Nādeyī' grows wild and its fruits are sour. [Verse 23]. Then comes the sweet tasting 'Kṣudra jambu' ('Madhura Kṣudrajambustu') which is also known as 'Kākajambu'. [Verse 24]. Fruits of *Psidium guajava* (Pārevataṁ) have a mixed taste of sweet and sour. The unripe fruit may be white and reddish in colour. It has a synonym 'Raivatakam'

Here ends the 'Jambukādi varga' [Verse 25].

Akṣapuṣpagaṇa

'Haritakī' (*Terminalia chebula*), 'Bibhitakī' (*Terminalia belerica*) 'Vātāma' (*Terminalia cattapa*) and 'Arjuna' (*Terminalia arjuna*) and those others with similar characteristics belong to the 'Akṣapuṣpagaṇa' (cf. Combretaceae). [Verse 26].

'Haritaki' has two synonyms 'Kaṣāya' and 'Haimavatī', and has astringent taste. Both fruits and seeds of 'Haritakī' have ridges on the surface ("Asrakphalabījañca"); the leaf is long ('Dīrghaparṇam'). Flowers of 'Bibhitaka' appear in spring ('Madhudrumah'). [Verse 27]. The fruit of 'Bibhiṭaka' (*T. belerica*) has the weight of a 'Karṣa' ('Karṣa' is a certain standard of weight). The fruit pulp of 'Bibhitaka' is astringent; the seed is brittle ('Karparabījaṁ'). The cotyledons contain oil ("Snighdhamajjā hi tailadah"). [Verse 28]. Vātāma bears new leaves in spring. Its fruit is shaped like an eye; the seed is brittle and contains oil. [Verse 29].

'Arjuna' (*T. arjuna*) has a synonym - 'Kakubha'. 'Arjuna' tree bears new leaves in spring and its bark is of red colour ("Vāsanto-aruṇavalkalaṁ"). It is also known as 'Dhavala'. Its seeds have angular ridges. The heart-wood of the tree is grey in colour. The fruit is such that its pulp and seed are inseparable: ("Dhumrasāraścabījaka"). [Verse 30].

Rarely the inflorescence of the Akṣapuṣpa bears sterile flowers. However when the flower is fertile the ovary is epigynous.

Here ends the 'Haritakyādi varga'. [Verse 31].

Kuhapuṣpa gaṇa

'Kola' and 'Badara' trees (family : Rhamnaceae) have prickles on their stems and branches ('Kaṇṭakadrumau'), and bear flowers which are 'Tungamaṇḍala' in nature (cf. Perigynous, vide Bījotpattikāṇḍa Ch. V. Verse 58). [Verse 32].

'Kola' or 'Badara' tree (Zizyphus jujuba), of the family Rhamanaceae has the synonym 'Sauvīraka'. The fruit is sweet. These trees grow naturally in arid regions ('Jāngolodhbhavaṁ', vide Bījotpattikāṇḍa, Ch. II, Verse 5). Badara fruits are sour and sweet and also slimy (Phalanca picchilaṁ bhavet). [Verse 33].

Here ends the 'Kolādi varga. Thus ends the first chapter of Vānaspatyavicaraṇādhyāya of the 'Vānspatyakāṇḍa' of the Vṛkṣāyurveda [Verse 34].

Saptaparṇādi Varga

Mallikā puṣpa gaṇa

'Saptaparṇa' (*Alstonia scholaris*) and 'Kuṭaja' (*Holarrhena antidysenterica* = H. pubescens) plants belong to the Mallikā puṣpa gaṇa (cf. Apocynaceae). Saptaparṇa is a big tree. The petioles are arranged in whorls, like the spokes of a wheel

(Arābandhanaṁ cf. Verticillate, vide Bījotpattikāṇḍa, Ch. IV, Verse 21). In the Kuṭaja plant the petioles are arranged successively in pairs (cf. opposite superposed vide Bījotpattikāṇḍa Ch. IV. Verse 18). Saptaparṇa and Kuṭaja plants blossom in autumn and during the rains respectively ("Śardi Saptaparṇaśca Kuṭajah Prāvṛṣyāṇe"). As they bear visible flowers, they come under the Vānaspatya kāṇḍa.

Mallikāpuṣpa has been already described in details in the Puṣpānga sūtrīyādhyāya (vide Bījotpattikāṇḍa Ch. V, Verse 73), and here I will describe the floral characteristics in short. In Mallikā puṣpa gaṇa, the flower is of 'Kuṇḍamaṇḍala' type (cf. Perigynous). Flowers in the inflorescence are arranged in an irregular manner. The calyx of Mallikāpuṣpa is gamosepalous with five incisions. It has gamopetalous corolla; the number of stamens is five. The two styles are united and produce Yugmaphala i.e., a pair of fruits. Plants of the Mallikāpuṣpa gaṇa (Apocynaceae) produce latex ('Sakṣīrā'). Fruits have many seeds. Seeds bear cotton like fibres ("Bahubījā ca tula-puccha samanwitā"). Saptaparṇa is a big tree; its latex is bitter. The tree is beautiful and can be grown as ornamental plants in the garden. It blossoms in autumn and its flower has a wine like smell ('Śārado madagandhakah'). Kuṭaja (*Holarrhena antidysenterica* = *H. Pubescens*) has four synonyms, viz. 'Vatsakah', 'Śakra', 'Kalinga' and 'Girimallikā'. (Note: The last two synonyms indicate that the Kuṭaja plants of the Mallikāpuṣpa gaṇa, abound in hilly tracts of Kalingadeśa that is Orissa). It is also known as 'Prāvṛṣya' (blossoms during the rainy season), 'Yavabīja', 'Tiktadru' (bitter bark and seeds) and 'Kṣīra pādapa' (Plants producing latex). The learned sages thus should be able to identify the plants of the Mallikā puṣpa gaṇa from the above description and names. Fig. 6a-b.

Sami Vargādhyāya

Parāśara said : I will now describe the 'Samivargādhyāya', dealing with the leguminous plants. [Verse 1]. 'Samivṛkṣa' - the leguminous plants bear 'Simbiphala' - the legumes; leaves are 'Simbiparṇa' or 'Punkhaparṇa ', in some other cases they bear 'Māhriparṇa' (For these three types of leaves-vide Bījotpattikāṇḍa, Ch. IV, Verses 53, 20 and 29 respectively). Flowers of the leguminous plants bloom during the day (Ravikānta, vide Bījotpattikāṇḍa, Ch. V, Verse 52). The 'Samivṛkṣa' grow mostly in dry arid climate. [Verse 2].

Sami plants (leguminous plants) are divided into the three sub families which has already been mentioned in chapter V on 'Puṣpāngasūtra' (Verse 65). The names and examples of these three subfamilies are given below, with their characteristics. [Verse 3].

They are 'Vakrapuṣpaṁ', 'Vikaraṇika puṣpaṁ and 'Śūkapuṣpaṁ. Now hear about the characteristics of the flowers of the three subfamilies of 'Sami Vṛkṣa'. [Verse 4]. At first I will describe the 'Vakrapuṣpaṁ'.

Vakrapuṣpaṁ (cf. Papilionaceae)

In 'Vakrapuṣpa' the flowers are heteromerous (Viṣamagaṇīya) with 'Tundamaṇḍala' insertion of floral leaves (cf. Hypogynous); gamosepalous flowers with five incisions; petals polypetalous and five in number, differently shaped (Viṣama) and bent ("Vakreṇa pañcakam bhavet"). Stamens ten of which nine are united. 'Simbiphala' (Legumes) have their seeds borne along the margin. ('Pārśwabījam). [Verse 5].

***Vikarṇikapuṣpaṁ* (cf. Caesalpiniaceae)**

In Vikarṇapuṣpamaṇḍal, insertion of floral leaves is of Tundamaṇḍal. (cf. Hypogynous) type. Sepals are five in number; gamosepalous; petals free of different shapes. Stamens are free and ten in number. [Verse 6]. Sometimes five stamens may be present. [Verse 7].

***Śūkapuṣpaṁ* (cf. Mimosaceae)**

Flowers of Śūkapuṣpa bear minute hairs. [Verse 8]. I will now describe each of the above subfamilies of the Leguminosae (Sami) separately; the characteristics will be described in Prākṛt (common language) while names (of the plants) will be mentioned in Sanskrit. [Verse 9].

'Palāśa' (*Butea monosperma*) has got several synonyms, viz. 'Kimśuka', 'Kṣāraśreṣṭha', (One of the best alkalies) 'Jāngala', Yājnika', 'Brahmapādapa'. It belongs to the family of 'Vakrapuṣpa' (cf. Papilionaceae) and its leaf is called 'Simbiparṇa' (cf. Compound leaf). [Verse 10].

'Agastya' (*Sesbania grandiflora*) bears leaf which is pinnate compound.– 'Punkhaparṇa.' It belongs to the subfamily of 'Vakrapuṣpa' (Papilionaceae). It has several synonyms like 'Munidrumah', 'Punkhaparṇī' and 'Vakrapuṣpikā'. [Verse 11-12].

'Jayanti' (*Sesbania aegyptica*) has synonyms, viz. 'Punkhaparṇī' (Leaves pinnate compound), 'Vakrapuṣpikā' (Flowers papilionaceous) and 'Nādeyī'. [Verse 12].

'Karanja' (*Pongamia pinnata*). which has synonyms like 'Naktamāl' and 'Putika' belongs to the subfamily of Vikarṇika (Caesalpinicaeae). [Verse 13].

'Śimśapā (*Dalbergia sissoo*) which is a big tree (with black heart-wood) and has two synonyms viz. 'Kṛṣṇasāra' and 'Mahādruma', belongs to the subfamily of 'Vakrapuṣpa' (Papilionaceae). [Verse 14].

'Pāribhadra' (*Erythrina indica*) belongs to the subfamily of 'Vakrapuṣpa' (cf. Papilionaceae). It bears compound leaves ('Simbiparṇa'). It is also called 'Sịmbataru'. [Verse 15].

'Āragvadha' (*Cassia fistula*) belongs to the subfamily of 'Vikarṇika puṣpa' (Caesalpiniaceae). [Verse 16]. Its flowers are yellow and the fruits are long and stick like called 'Kapidaṇḍa'. It has several synonyms, viz. 'Rāja Vṛkṣa', 'Swarṇāṭu', 'Caturangula', 'Sampāk' and 'Kṛtamāla'. The leaf is pinnate compound ('Punkhaparṇa'). [Verse 17].

'Kovidāra' (*Bauhinia purpuria*) is a plant belonging to the subfamily of Vikarṇapuṣpa (Caesalpiniaceae). Flowers are white. The leaf is a 'Yugmapatra' (vide Bījotpattikāṇḍa, Ch. IV, Verse 54). Sepals and petals are five; five stamens are free. [Verse 18]. Fig. 6c.

'Kāñcanāra' (*Bauhinia variegata*) belongs to the sub family Vikarṇika (Caesalpiniaceae). The leaf is a 'Yugmapatra'. [Verse 19].

'Aśmantaka' also belongs to the subfamily, Vikarṇika puṣpa. Its flowers are yellow. The bark is fibrous ('Angśuka, vide Bījotpattikāṇḍa Ch. VII, Verse 17) and it is soft ('Ślakṣṇavalkalaṁ). It is called 'Pūta Vṛkṣa'. [Verse 20-21].

'Śirīṣa' (*Albizzia lebbeck*) belongs to the subfamily, 'Śūkapuṣpa' (Mimosaceae). The leaf is a 'Punkhaparṇa'(Pinnate compound) . It has several synonyms viz. 'Śukapriya', (Note: A kind of bird - 'Śuka', prefers living in Śirīṣa tree). 'Udyānaka', 'Dṛdhasāra', 'Śyāmala', 'Simbinīphala'. [Verse 22]. Fig. 6d.

'Khadira' (*Acacia catechu)* belongs to the subfamily 'Śukapuṣpa' (Mimosaceae). The leaf is small and round (Bālapatra, vide Bījopattikāṇḍa Ch. IV, Verse 55); leaves are pinnate compound ('Punkhabandhana' type - vide Bījotpattikāṇḍa, Ch. IV, Verse 20). The heart-wood is reddish (Raktasāraśca). The tree has prickles. Wood is used in making sacrificial fire ("Yājnikah Srṃtah"). It has a synonym - 'Gāyatrī'. [Verse 23]. 'Kadara' is another species of 'Khadira'. It has two other synonyms viz. Somavalka and Pathidrumah. [Verse 24].

'Sami' (*Mimosa suma*) plants bear 'Bālapatra' (Small round leaves). Leaves are of 'Punkhaparṇī' type. 'Sami' plant is the best material for kindling the fire of Vedic rites ('Samidwarah'). Thus 'Sami' is a sacred tree (Pavitra). The leaf buds of 'Sami' are called 'Śanku-Kalikā' (shaped like a dagger, vide Bījotpattikāṇḍa Ch. IV 45). [Verse 25].

'Ari' (*Acacia farnesiana*) is another species of subfamily Śūkapuṣpī (Mimosaceae). It bears yellow flowers with fragrance ("Pītapuṣpī sugandhikā"). The leaf is of 'Punkhaparṇa' type; leaves are very small and the trees flourish in deserts. ("Bālapatri ca maruja"). [Verse 26].

'Viṭkhadira' (synonym 'Kāmbhojī') plant has got bad smell ('Durgandhapuṣpikā') and its flower has yellow colour. The plant Khadira and its species have astringent taste, and prickles all over their body. [Verse 27].

Śirīṣa and its species (*Albizzia* species) - all belong to the subfamily Śūkapuṣpa (Mimosaceae). [Verse 28].

'Aśoka' (*Saraca indica*) tree has a synonym - 'Vanjula'. It bears fruits characteristics of leguminoseae plant. Flowers are borne in the inflorescence in a cluster ('Guccha mañjari'). They appear in spring; the new leaves are red ("Vāsanto raktapallavah"). [Verse 29].

'Śyonāka' (*Oroxylum indicum*) is a leguminous plant having a long fruit ('Dīrgha simbi'). It has two synonyms : 'Vanjula' and 'Bhūtapuṣpa'. Its leaf is shaped like the outline of an eye (Mañjuparṇakaṁ - vide Bījotpattikāṇḍa Ch. IV, Verse 38). Śyonāka leaves (leaf - lets) are borne on the rachis (·Mārhi). Stem or bark is pungent; the bark is yellowsih. [Verse 30].

'Tintirī' (*Tamarindus indica*) is a leguminous tree of mixed characters. It bears characteristic flowers. Its fruit is sour. It has three synonyms, viz. 'Amlikā', 'Vṛkṣāmla' and 'Kṣārpādapa'. (Note : The last one indicates that ash of the tree is alkaline). [Verse 31].

वृक्षायुर्वेदः ।

गुल्मक्षुपकाण्डः ।

दल्वजवर्गः ।

अथातो दल्वजवर्गाध्यायं व्याख्यास्याम इत्याह पराशरः। इह खल्वानुपदल्वच्छन्न भूमिषु सञ्जातमुदभिदं दल्वजं विज्ञायते । तेषां दल्वजानामेकमातृकं कन्दरोहकं त्रिसिराकमठच्छदं यदुद्भिदं तदेव कण्डुला संज्ञायते । एव तासां दीर्घं नालसङ्काशं पत्रवृन्तमन्तरसुषिरञ्च । तस्मिन् बहुनि स्रोतांसि विद्यन्ते । वृन्ते पत्रसन्निवेशं पृष्ठग्रन्थिकञ्च। एष वर्गोऽक्षमञ्जरिकश्च। एव तासामेकमोचिकोतुपुच्छिकवल्लर्यां सफलं निष्फलञ्च द्विविधं पुष्पं भवति । कण्डुलानां स्वरसः किञ्चिद्विषविकाशिगुणत्वात् स्पर्शे कण्डुश्च भवति । कण्डुला प्रकृत्या पञ्चविधा भवति । मानका च ग्रन्थिला च खण्डकर्णा च वल्लीकण्डुला चैकपर्णिका चेति । तत्र मानका वृहत्कमठच्छदं भवति । गते बहुवर्षे कदाचिद हरितमोचिकेन पुष्पमञ्जरिर्दृश्यते । मानका तु त्रिविधा भवति। निर्विषा च सविषा च क्षुद्रमानकेति । तत्र निर्विषा बृहत्पत्रं मध्यमकन्दं भवति । सविषा तु बृहत्कन्दं बहुवर्षं जीवयति । क्षुद्रमानकैव सविषा भवत्येव वार्षिका च । ग्रन्थिला तु पुष्पान्ता भवत्येव मूलेन ग्रन्थिः सञ्जायते । अस्याः पीतमोचिकेन सह पुष्पमञ्जरिर्दृश्यते । खण्डकर्णा तु मण्डुकपर्णिनी भवति । अस्या हरितमोचिकेन सह रक्तमञ्जरिर्दृश्यते। एव फलपाकान्ता भवति । बीजेन वा कन्देन प्ररोहयेत । वल्लीकण्डुला तु प्रताना स्यात् । अस्यां खलु पत्रमञ्जरिभेदेन बहुभेदो विद्यते। एकपर्णिकश्च वर्तुलकन्दश्च । स तु सुरणकन्दमाचक्षते ।

अस्यादौ पुष्पमञ्जरिः सम्भवति । पश्चात् शुङ्गावृतं दीर्घनालवृन्ते पृष्ठग्रन्थिकमेकपर्णं सञ्जायते । एव यावज्जीवेत्तावदेकपर्णेन विद्यते । पर्णन्तु बहुकर्णं भवति । इति दल्वजवर्गे कण्डुलागणाः ।

वृक्षायुर्वेदः ।

गुल्मक्षुपकाण्डः ।

अथ गुल्मान् प्रवक्ष्यामि सिंहपुष्पेण भाषितान् ।
समपक्षं भवेत् पर्णं दीर्घेण यच्च संज्ञितम् ।।
सिंहपुष्पं जृम्भदलं विषमकुण्डमण्डलम् ।
पलाशवल्लिरिस्तत्र व्यत्यासपक्षपङ्क्तिकम् ।।
पञ्चकर्ण भवेद्दलं कुण्डाङ्ककेशरस्तथा ।
द्विकेशरौ हयक्रान्तौ द्विवराटौ द्विवर्तकौ ।।
कदाचित् सिंहपुष्पञ्च भिद्यते केशरेण च ।
चत्वारः केशरास्तत्र विषमेण भवन्ति हि ।।

वृक्षायुर्वेदः ।

विरुधवल्लीकाण्डः ।

अथवल्लीतृणगण्डालिकावर्गः ।

खलु वल्लीतृणगण्डालिका विविधाः सन्ति । एता मिश्रतृणजातयः । प्ररोहे पर्णामेकं भवति । एषां वर्गाणां पत्राणि मञ्जुकसंज्ञकानि स्युः । पत्रसिराणां संस्थानं वेल्लितञ्च। पत्रबन्धनन्तु पट्टिकेन काण्डक्रान्तं भवति । काण्डस्तु तन्तुलं बहुसन्धितञ्च भवति । स्वल्पप्रताना एव तृणपुष्पेण विद्यात् । स्वरसस्तु सान्द्रं पिच्छिलं भवति । एष गणः काञ्चलिका नामतो विज्ञायते गण्डिनी वा । एव पत्रेण पुष्पेण विविधं भवतीति तृणगण्डालिकावर्गः ।

विरुधवल्लीकाण्डः ।

(१) सूर्यवल्ली, (२) चन्द्रवल्ली, (३) सोमवल्ली, (४) भद्रवल्ली, (५) कन्दवल्ली, (६) काष्ठवल्ली, (७) कण्टवल्ली, (८) छिन्नरुहावल्ली, (९)

त्रिपुटावल्ली, (१०) कोलवल्ली, (११) तुम्बीत्रिपुषवल्ली, (१२) जालिनीकृतबेधनवल्ली, (१३) आकाशवल्ली, (१४) क्षीरवल्ली, (१५) गुच्छफलवल्ली, (१६) सिम्बिफलवल्ली (१७) मञ्जुवल्ली। (१८) जलवल्ली, (१९) दल्वजवल्ली, (२०) ह्रस्ववल्ली, (२१) तृणगण्डालिकावल्ली, (२२) ऋतुवल्ली, (२३) कपोतवल्ली ।

Gulmakṣupa Kāṇḍa*
Dalwajavargādhyāya

Parāśara said : I will now describe the chapter on the plant group - 'Dalwajavarga'. These are plants which abound in a land that ever remains moist and clayey. The plants of this 'Varga' have mostly monocotyledonous seeds; plants are propagated by rhizomes ('Kandarohakaṁ'). The leaves of these plants are shaped like the outline of a tortoise ('Kamaṭhacchadam') and have three main veins ('Trisirā'). These plants are called 'Kaṇḍulā' (Note: 'Kaṇḍulā' means that which causes itching). 'Kaṇḍulā' plants have long, tubular petioles with innumerable vascular channels (Srota). The petioles of these plants are attached to the back of the leaf lamina ('Pṛsthagranthikanca'). 'Kaṇḍulā' group of plants bear 'Akṣamañjari' (vide Bījotpattikāṇḍa Ch. V. Verse 16) type of inflorescence having only one spadix which looks like the tail of a cat ('Otupucchikā vallaryam'). Flowers are both fertile and sterile.

The sap of 'Kaṇḍulā' plants are irritating to the skin; naturally, contact produces itching. 'Kaṇḍulā' plants are of five kinds, viz., 'Mānaka', 'Granthilā', 'Khaṇḍakarṇa', 'Vallīkaṇḍulā' and 'Ekaparṇī'. The leaf of 'Mānaka' (*Alocasia indica*) looks like a big tortoise (the outline shape). The inflorescence of 'Mānaka' comes out after many years

---

* Verse number was found missing.

along with a green spathe. Mānaka plants are of three kinds i) nonirritating, ii) irritating and iii) 'Kṣudra Mānaka' (*Colocasia antiquorum*). The nonpoisonous or non-irritating Mānaka has a medium sized rhizome with big leaves whereas the poisonous variety has big rhizomes, and lives for many years. Kṣudramānaka, species of *Colocasia* is a poisonous plant that dies within a year. The 'Granthilā' variety dies whenever its flowers appear, but buds grow again from the root. Its inflorescence has a yellow coloured spathe. There is another variety of *Colocasia* sp. whose leaf is partially incised ('Khaṇḍakarṇa') and is shaped like a frog, ('Maṇḍukaparṇī', vide Bījotpattikāṇḍa Ch. IV, Verse 67). The inflorescence of this variety is red ('Raktamañjari'); bears green coloured spathe ('Mocika'). They are annuals and are propagated by seed or rhizome. 'Vallīkaṇḍulā' produces runners (e.g. *Colocasia antiquorum*). These are of many kinds according to their inflorescences ("Patramañjaribhedena"). 'Suraṇakandaṁ' (*Amorphohallus campanulatus*) has only one stem with one leaf; its underground corm is round. It has many scaly leaves ('Śungāvṛtaṁ'). The long stem with innumerable tubular passages holds only one peltate leaf. The plant bears only one stem with a leaf that has many incisions.

Here ends the Kaṇḍulā gaṇa of Dalwajavargādhyāya.

Gulma Kṣupakāṇḍa

Parāśara said : I will now describe the group 'gulmas' or 'Kṣupas' (Shrubs) with reference to the 'Simhapuṣpa', (cf. Acanthaceae). In plants of the 'Simhapuśpa type, the two halves of the leaf blade are of equal size, and the leaf is somewhat longer. Fig. 7.

The flowers of the plant known as 'Simhapuśpa' have corolla which resembles a "widely opened mouth of a lion" (Vide Bījotpattikāṇḍa Ch. V. Verse 34). The petals of 'Simhapuśpa' are of unequal size and flowers are epigynous.

The plant of the Simhapuśpa gaṇa (cf. Acanthaceae) bears inflorescence which is of 'Palāśvallari' type (vide Bījotpattikāṇḍa Ch. V, Verse 9). Petals of the flower of Simhapuṣpa are five in number; stamens are epipetalous ('Kuṇḍankakeśara'); stamens are two; exceptionally they may be four of unequal size. The shape of the stamen is of 'Hayakrānta' (shape of a horse's hoof) type. The ovary has two styles, and two locules.

Virudha Vallī Kāṇḍa*
Vallī Tṛṇagaṇḍālikā Varga

What Parāśara means by 'Vallī-Tṛṇagaṇḍālikā' cannot be fully clarified. However, Parāśara said : there are many plants in this varga. They are a mixed type of grass like plants ("Eta miśra tṛṇajātaya"). On sprouting they produce only one leaf. Their leaves are termed 'Mañjuk' (Parallel Venation), may have reticulate venation also ('Vellitanca'). The leaf ensheathes the stem (Paṭṭikaṁ). The stem is creeping in nature; it is fibrous and has many segments. Flowers are like that of the grass family ("Tṛṇapuṣpeṇa vidyāte"). The sap of these plants are thick and slimy. Plants of this family are called 'Kāñcalikā' and also 'Gaṇḍinī'. They have got different species characterised by the form of their leaves and flowers.

In Virudhavallī Kāṇḍa Parāśara has mentioned the following plants:

(1) 'Sūrya vallī', (2) 'Candra vallī', (3) Soma vallī', (4) 'Bhadra vallī', (5) 'Kanda vallī, (6) 'Kāṣṭha vallī', (7) 'Kaṇṭa vallī', (11) 'Tumbītripuṣavallī' (12) 'Jālinīkṛtabedhana vallī, (13) 'Ākāś vallī.', (14) Kṣīra vallī, (15) 'Gucchaphala vallī', (16) 'Simbiphala vallī, (17) 'Mañjuvallī' (18) 'Jalavallī', (19) 'Dalwaja vallī, (20) 'Hraswavallī', (21) 'Tṛṇagaṇḍālikāvallī, (22) 'Ṛtuvallī', (23) 'Kapotvallī.

---

* Verse number was found missing in the manuscript.

Most of the above mentioned plants of the 'Virudhavallī' kāṇḍa have been identified, through cross references to some of the Āyurvedic pharmacopoeias, viz. Pātanjal Nighanṭu', 'Rāja Nighanṭu', 'Madanpāl Nighanṭu' etc., and noted as under:

1. Sūrya vallī could not be identified.
2. Candra vallī Sans. Mādhavi (*Hiptage benghalensis* kurtz. syn. *H. madablota* Gaertn., Malpighiaceae).
3. Soma vallī -could not be identified.
4. Bhadra vallī Sans. Praśāraṇī (*Paederia foetida* L., Rubiaceae).
5. Kanda vallī Sans. Vidari kanda (*Ipomoea paniculata* R. Br. Convolvulaceae).
6. Kāṣṭha vallī Sans. Dāruharidrā (*Berberis aristata* DC - Berberidaceae) Fig. 8a.
7. Kaṇṭa vallī Sans. 'Śatamūli or Satavari (*Asparagus racemosus* Willd. Liliceae).
8. Cchinnaruhā vallī Sans. Amṛta vallī, Syn Guduci or Cchinnaruhā (*Tinospora cordifolia* Willd. Menispermaceae). Fig. 8b.
9. Tripuṭa vallī Sans. Tripuṭa or Trivṛt (*Operculina turpethum* L. Convolvulaceae).
10. Kola vallī - Kola valli includes some plants of the Piperaceae family, viz. *Piper betle* L., *Piper chaba* Hunter, *Piper longum L. Piper nigrum* L. The Sanskrit names are 'Tambuli', 'Cavya', 'Pippali' and 'Marica' respectively.
11. Tumbī Tripuṣa vallī covers all plants of Cucurbitaceae family.
12. Jālinīkṛtabedhana vallī: Plants of this group belong to the Cucurbetaceae family; their fruits when ripe develop a kind of fibrous and rough skeleton.
13. Ākāś vallī (*Cassytha filiformis* L.).
14. *Kṣīra* vallī (*Ipomoea panciulata* R.Br.).

15. Gucchaphala vallī Sans. Drākṣā plant; Grape vine (*Vitis vinifera* L. Vitaceae).
16. Simbiphala vallī Sans. Yaṣṭhimadhu (*Glycyrrhiza glabra* Leguminosae, vide : Samivargādhyāya of Vānsaspatya Kāṇḍa Ch. II, Verse 22).
17. Mañju vallī Sans. manjiṣṭhā, syn. Netraparṇinī and Raktanāla (*Rubia cordifolia L. Rubiaceae*)
18. Jala vallī. Sans. Kalambi, (*Ipomoea aquatica* Forsk. Convolvulaceae).
19. Dalwaja vallī: Possibly meaning all sorts of creepers which grow in 'Dalwa bhūmi' that is moist and clayey land (vide : Bījotpatti Kāṇḍa, Ch. II, Verse 16).
20. Hraswa vallī - could not be identified.
21. Tṛṇagaṇḍālikā vallī - could not be identified.
22. Ṛtu vallī - could not be identified.
23. Kapota vallī - could not be identified.

Here ends the Virudha valli kāṇḍa the of Vṛkṣāyurveda By Parāśara.

Bibliography

Agnipurāṇa -Edited by Pandits of Ānandaśrama Press, Poona, 1822.

Amarkośa (1) Translated into English by H.T. Colebrooke, Serampur, 1808.
(2) Edited and traslated into Bengali by Gurunath Bhattacharya, Sanskrit Pustak Bhander, Calcutta, 1990.

Arthaśāstra -Translated into English by R. Shamasastry with an introductory note by J.F. Fleet, 4th edition, Mysore, 1951.

Atharvaveda -Translated into English by R.T.H. Griffith, 2 vols., Chowkhamba Sanskrit Series Office, Varanasi, 1968.

Benerji, S.C. -Flora and fauna in Sanskrit literature, Naya Prakash, Calcutta 1980.

Bose, D.M., Sen, S.N. and Subbarayappa, B.V. -A concise history of Science in India. Indian National Science Academy, New Delhi, 1971.

Bose J.C. -Nervous Mechanism of plants (Lognman) 1926.

Bṛhatsaṃhitā -Edited with English translation by V. Subramanya Śāstri and M.R. Bhat, 2 vols., Bangalore 1947.

Caraka Samhitā - Edited with English, Hindi, Gujarati translations, 6 vols, Gulab Kunverba Āyurvedic Society, Jamnagar, 1949.

Chāndogya Upaniṣad -Edited by Ranganath Śāstri Vaidya with the commentary Mitaksara of Nityananda, Poona, 1915.

Chopta, R.N., Nayar, S.L. and Chopra I.C. -Glossary of Indian Medicinal Plants, CSIR, New Delhi, 1956.

Dīgha Nikāya -Edited by T.W. Rhys Davids and J.E. Carpenter, Pali Text Society, 3 vols., London, 1890.

Gangulee, H.C., Das, K.S. and Datta, C. -College Botany, vol. 1, New Central Book Agency, Calcutta, 1972.

Gopal, Lallanji -Aspect of History of Agriculture in Ancient India, Bharati Prakashan, Varanasi, 1980.

Kāmasūtra of Vātsyāyana -Edited by Damodara Śāstri Gosvami with the commentary jayamangala of Yaśodhara, Benaras, 1929.

Madanpāla Nighaṇṭu -Edited by Vaidya Bhagwan Dash and Assisted by Kanchan Gupta, B. Jain Publishers, Delhi, 1991.

Mahābhārata -Edited by V.S. Sukthankar and others, 22 vols., Bhandarkar Oriental Research Institute, Poona, 1933-59.

Majumdar, G.P. -Vanaspati, Calcutta University, 1927.

Majumdar, G.P. -Upavana-Vinoda, the Indian Research Institute, Calcutta, 1935.

Majumdar, G.P. -The History of Botany and Allied Sciences in Ancient India, AIHS, UNESCO, Paris, 1951.

Mānasollāsa of Someśvara -Edited by G.K. Shrigondekar, Baroda, 3 vols., Gaekwad's Oriental Series, 1925, 1939, 1961.

Manu Samhitā -Edited by Ganganatha Jha, BI, No. 256, 3 vols., Asiatic Society, Calcutta, 1936-39.

Prain, D. -Bengal Plants, I-II, Calcutta (Reprinted edition 1963, Botanical Survey of India, Calutta).

Rājanighaṇṭu and Dhanvantari Nighaṇṭu -Edited by Hari Narayan Apte, Ānandāśrama press, Poona, 1818.

Seal, B.N. -Positive Sciences or Ancient Hindus, London, 1915. Reprinted by Motilal Banarsidass, 1985.

Sircar, D.C. - Studies in the Geography of Ancient India and Medieval India, 1971.

Sircar, Nityendra Nath-An introduction to the Vṛkṣāyurveda of Parāśara, JASB, 16, 123-139,1950.

Suśruta Samhitā -Translated into English by Kunjalal Bhiṣāagratna, 3 vols., Calcutta, 1907.

Taittirīya Samhitā -Edited by Roer and Cowell, 6 vols., Calcutta, 1854-99. Translated by A.B. Keith, HOS, Nos. 18. 19, 1914.

Vājasaneyī Samhitā or Śukla Yajurveda Samhitā -Edited by Vasudeva Laksmana Śāstri Panasikara, Bombay, 1929. Translated by R.T.H. Griffith, 1899.

Vṛkṣāyurveda of Śūrapāla -Ms. No. 324D (Mr. Walker, 187), Bodleian Library, Oxford.

Index

[Alternative sources are indicated within parenthesis. Abbreviations : A.K.- Amarkośa; B.P. - Bhāva Prakās; C.S. - Caraka Samhitā; D.N. - Dhanvantari Nighaṇṭu; R.N. - Rājanighaṇṭu; S.S. - Suśruta Samhitā]

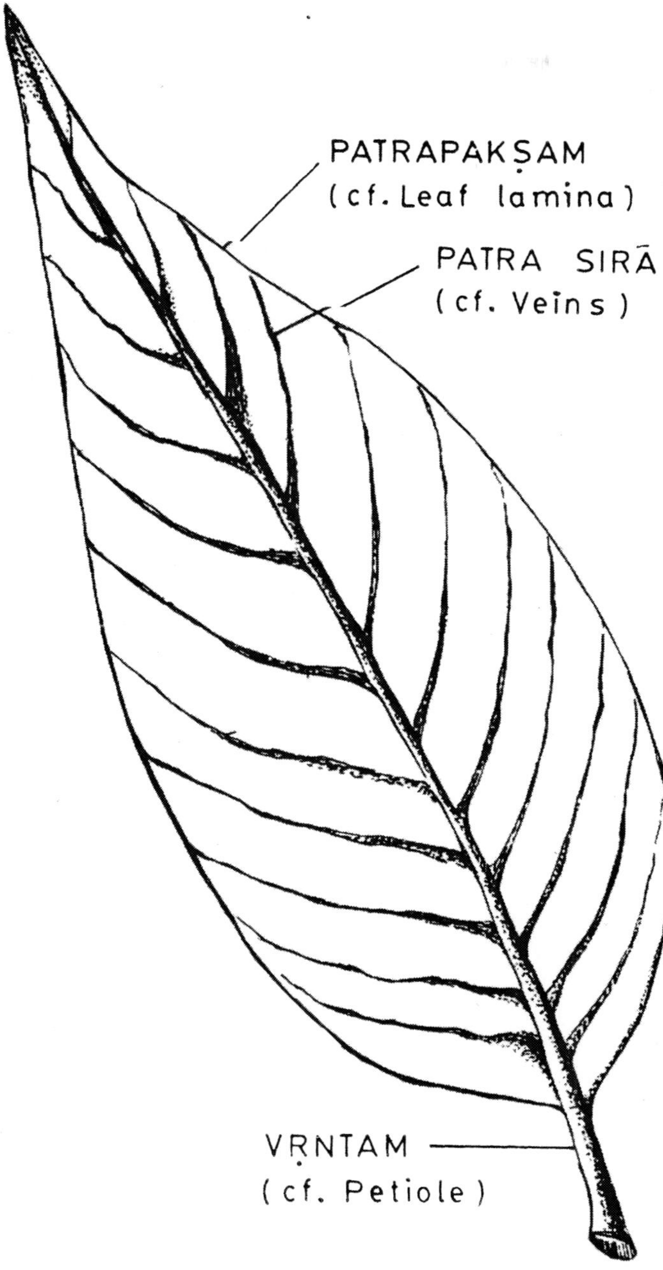

Fig. 1 : Parts of a leaf

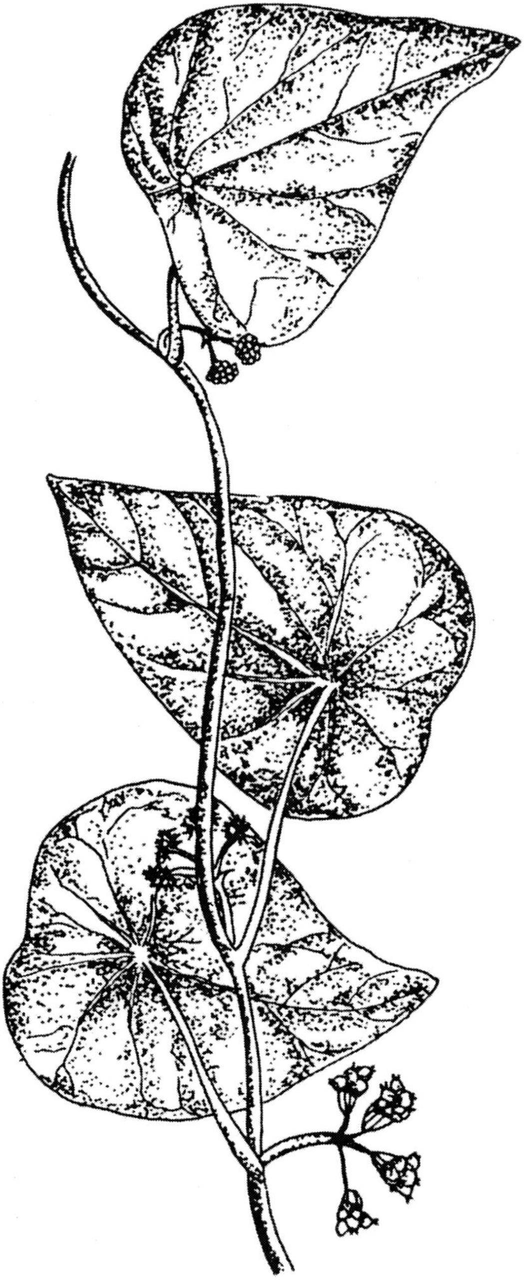

Fig. 2 : (a) Arghya patra
(e.g. Leaf of *Stephania hernandifolia* Willd Walp)

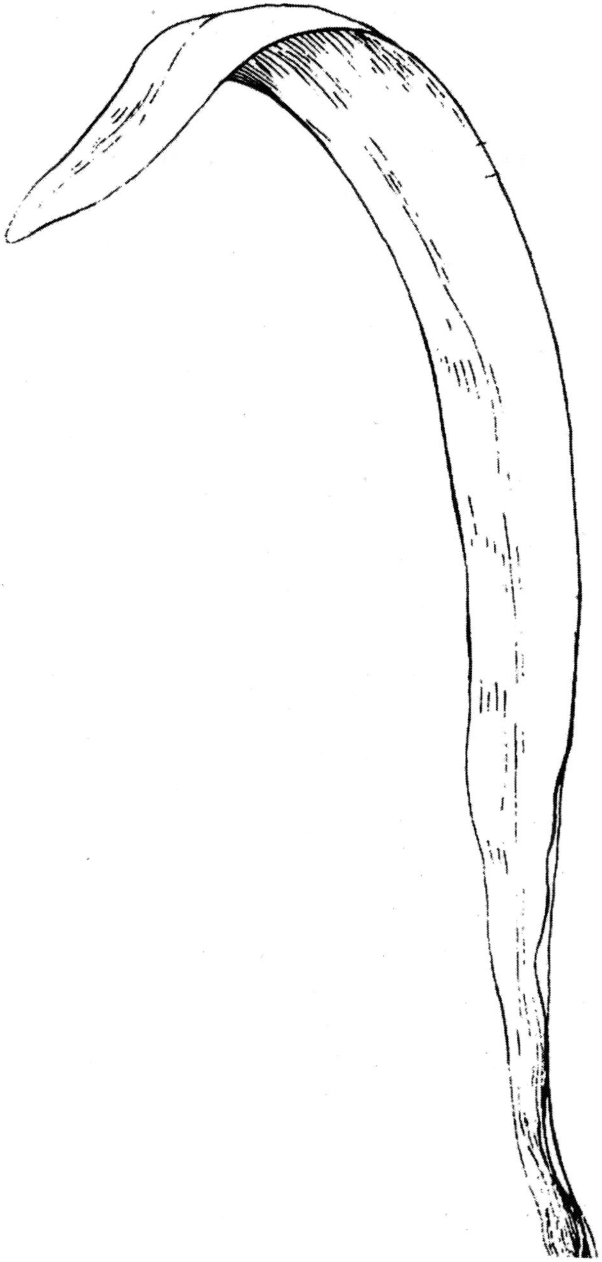

Fig. 2 : (b) Vīrapatra
(cf. Lanceolate, e.g. Leaf of *Lilium* sp.)

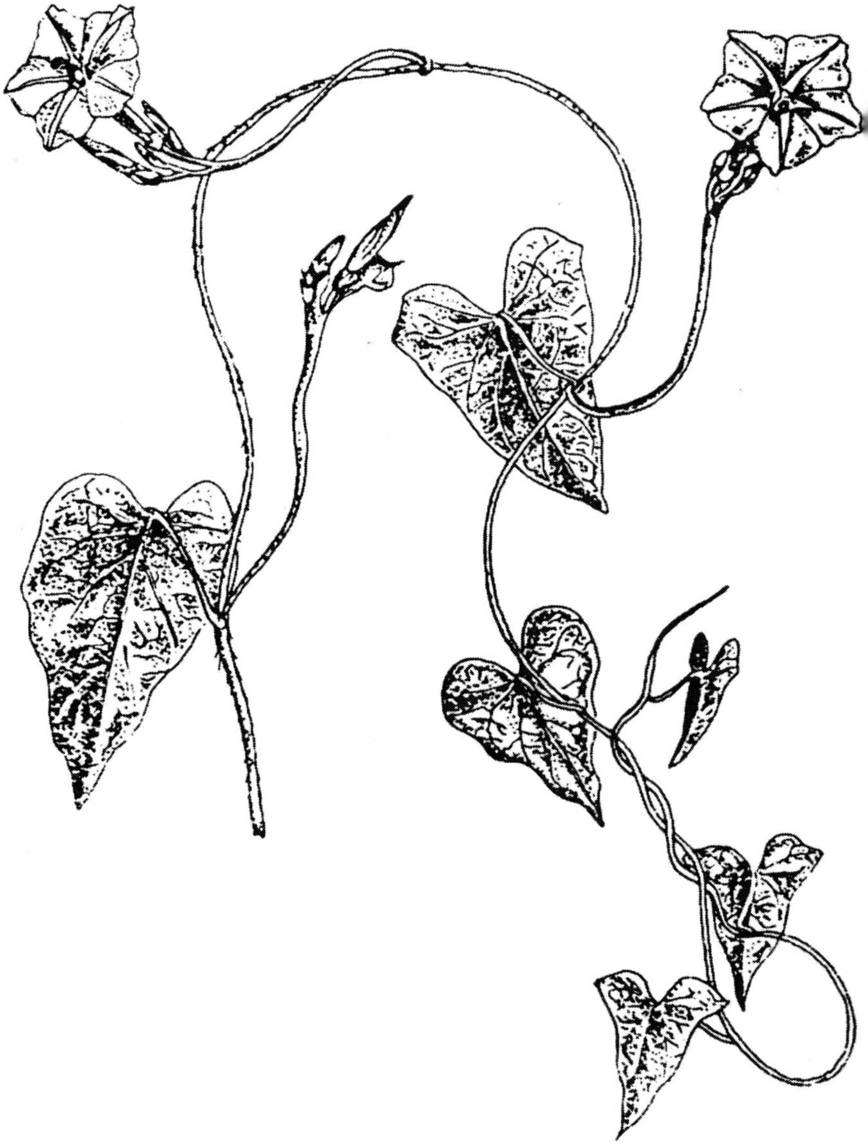

Fig. 2 : (c) Vāṇavarha
(cf. Hastate; e.g. Leaf of *Ipomoea* sp.)

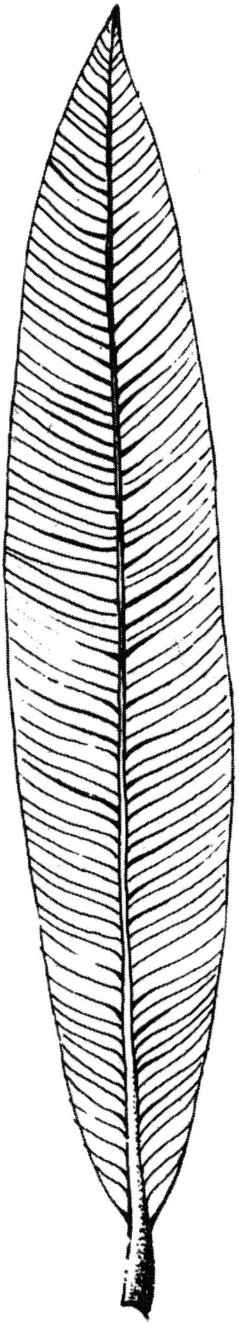

(A leaf with the shape of a knife e.g. *Nerium oleander* L.)

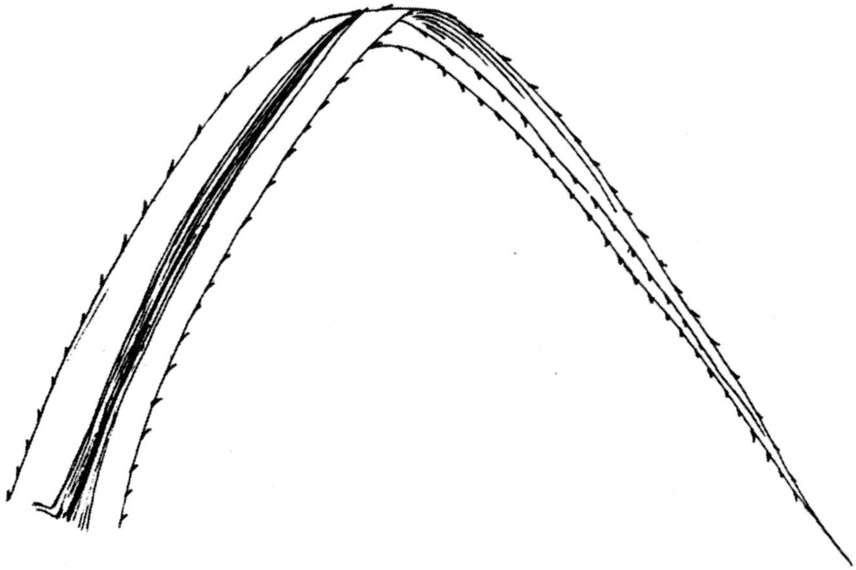

Fig. 2 : (e) Krakacchada
(cf. Serrate, e.g. *Pandanus sp*)

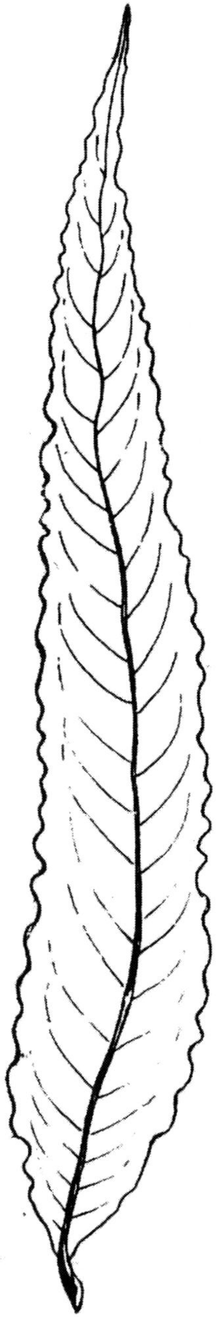

Fig. 2 : (f) Vṛjina patra
(cf. Repand, e.g. Leaf of *Polyalthia longifolia* Benth & Hook. f)

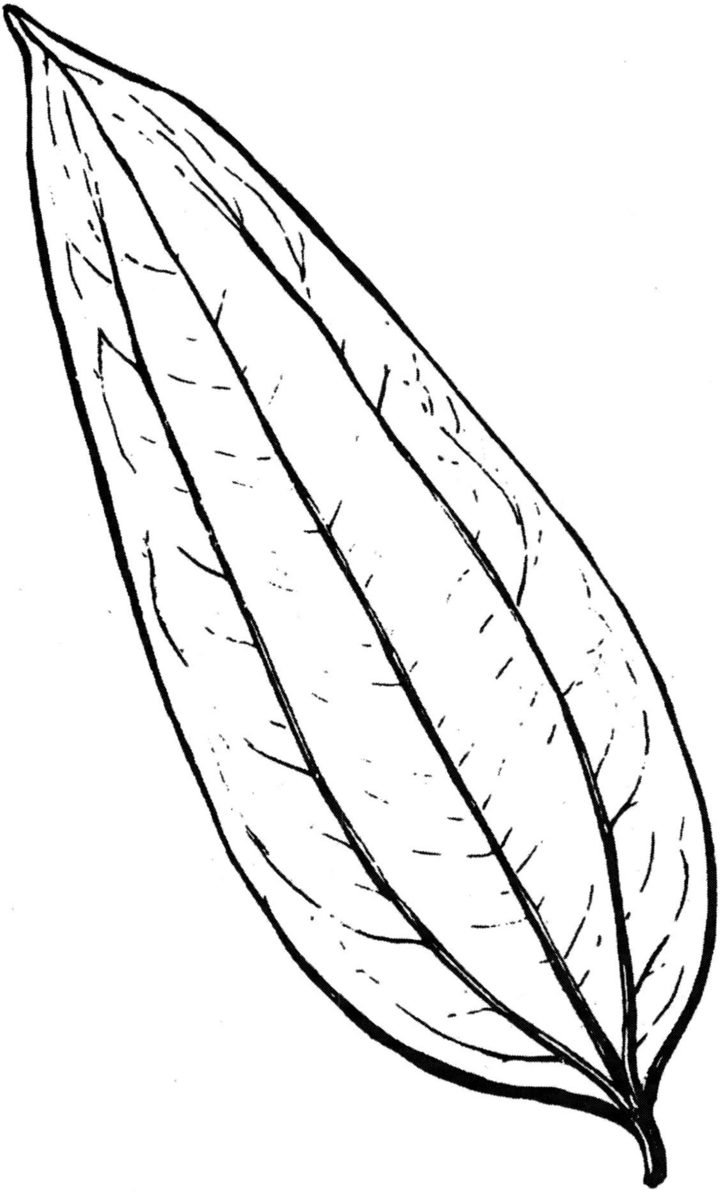

Fig. 2 : (g) Mañjuvarha
(A leaf shaped like an eye, e.g. *Cinnamomum obtusifolium* Nees)

Fig. 2 : (h)Citraparṇa
(A leaf with surface adorned with colourful spots e.g. *Croton* sp)

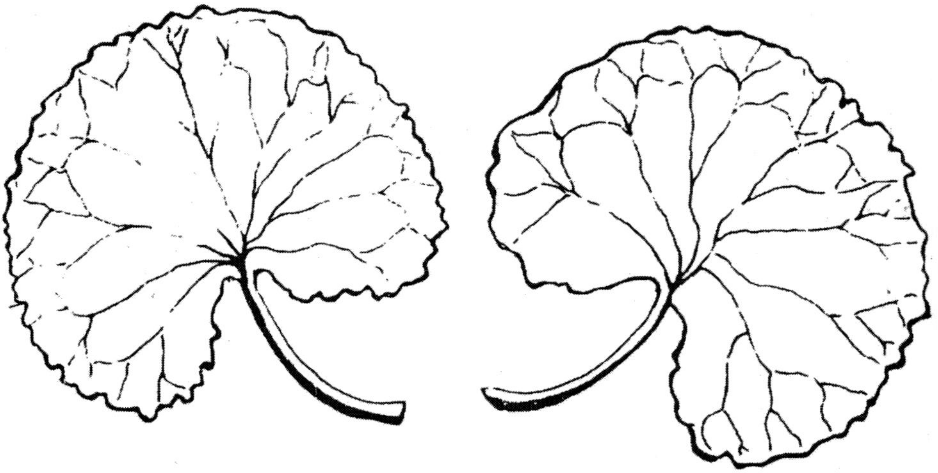

Fig. 2 : (i) Maṇḍukaparṇa
(cf. Reniform leaf, e.g. ***Centella asiatica*** **L.)**

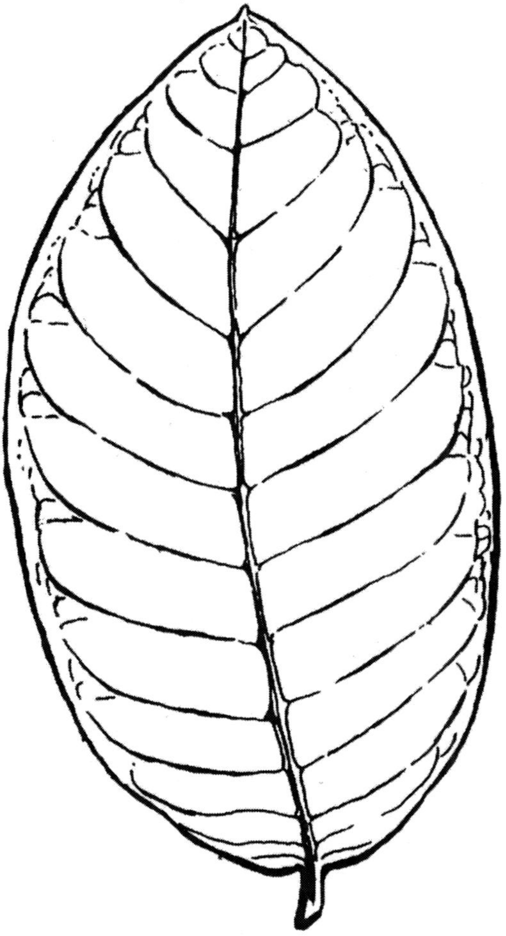

Fig. 2 : (j) Śuktiparṇa
(cf. an elliptical leaf, e.g. *Psidium guajava* L.)

Fig. 2 : (k) Hastiparṇī
(e.g. Leaf of *Abroma augusta* L.)

Fig. 2 : (I) Puṣpapatraṁ
(cf. Bract, e.g. *Poinsettia pulcherrima*)

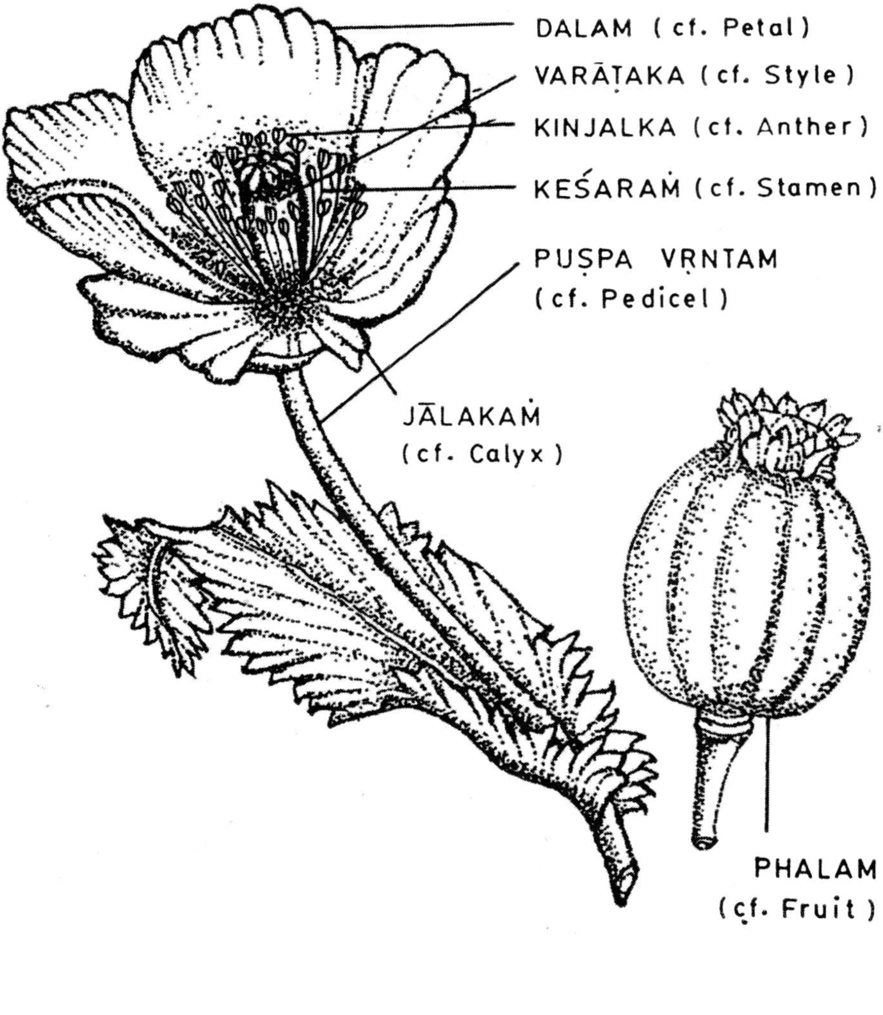

Fig. 3 : Parts of a flower

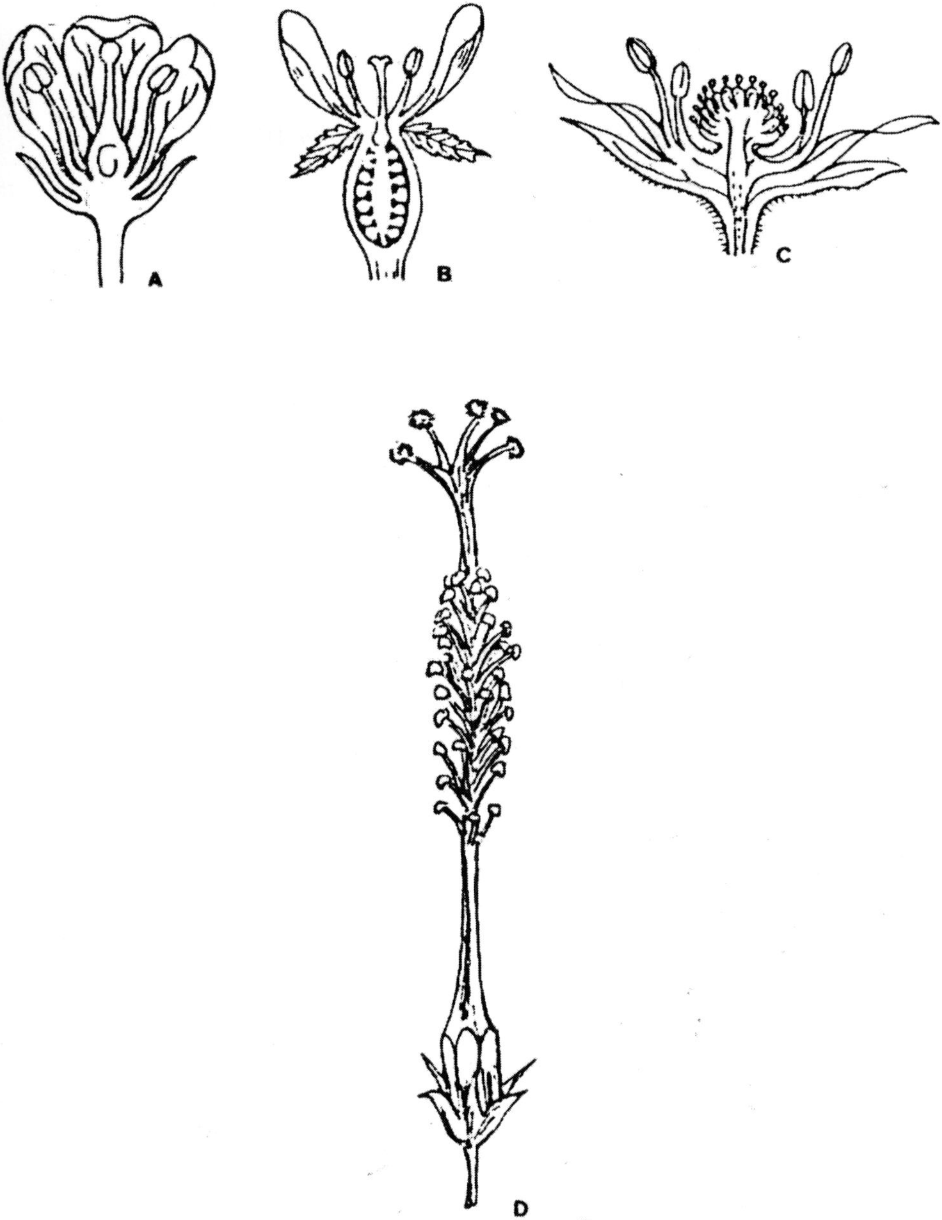

Fig. 4 : Insertion of floral leaves on the thalamus
(a) Tundamaṇḍala (cf. Hypogynous)
(b) Kumbhamaṇḍala (cf. Epigynous)
(c) Tungamaṇḍala (cf. Perigynous)
(d) Vāṭyamaṇḍala

Fig. 5 : (a) Vaṭa (*Ficus bengalensis* L.)

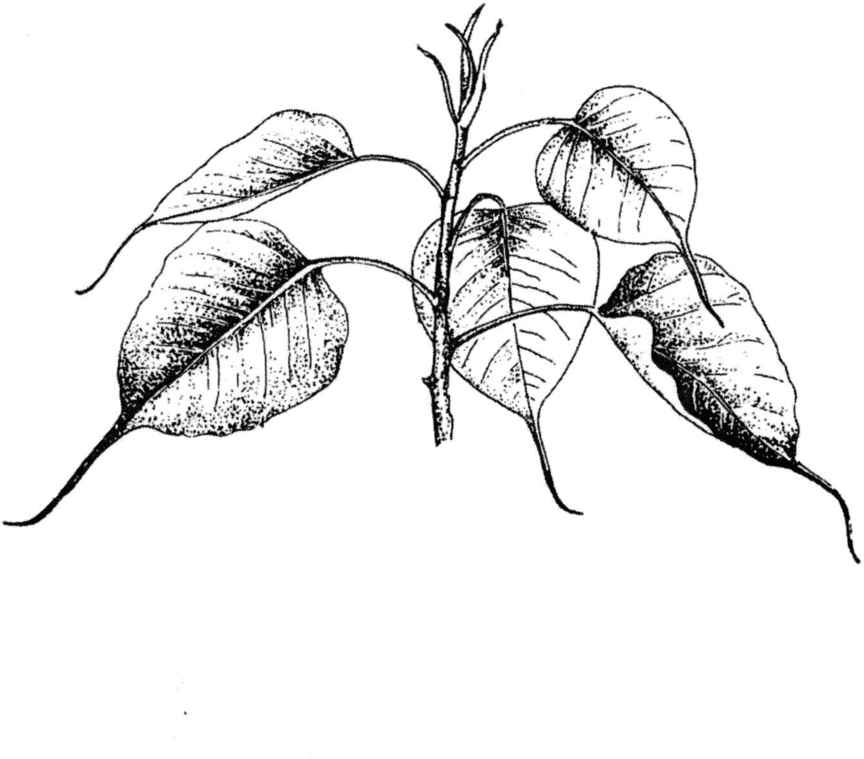

Fig. 5 : (b) Aśwattha (*Ficus religiosa* L.)

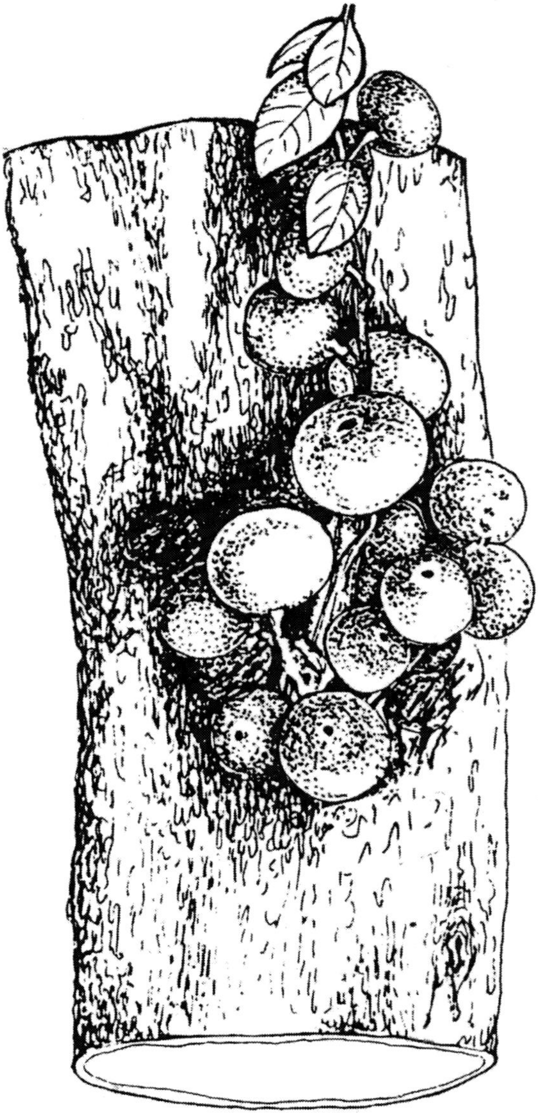

Fig. 5 : (c) Yajñoudumbara (*Ficus glomerata* Roxb.)
showing 'Kāṇḍa Phala' (fruits formed on the stem itself)

Fig. 5 : (d) Panasa (*Artocarpus heterophyllus* Lam.) showing 'Kāṇḍa Phala'.

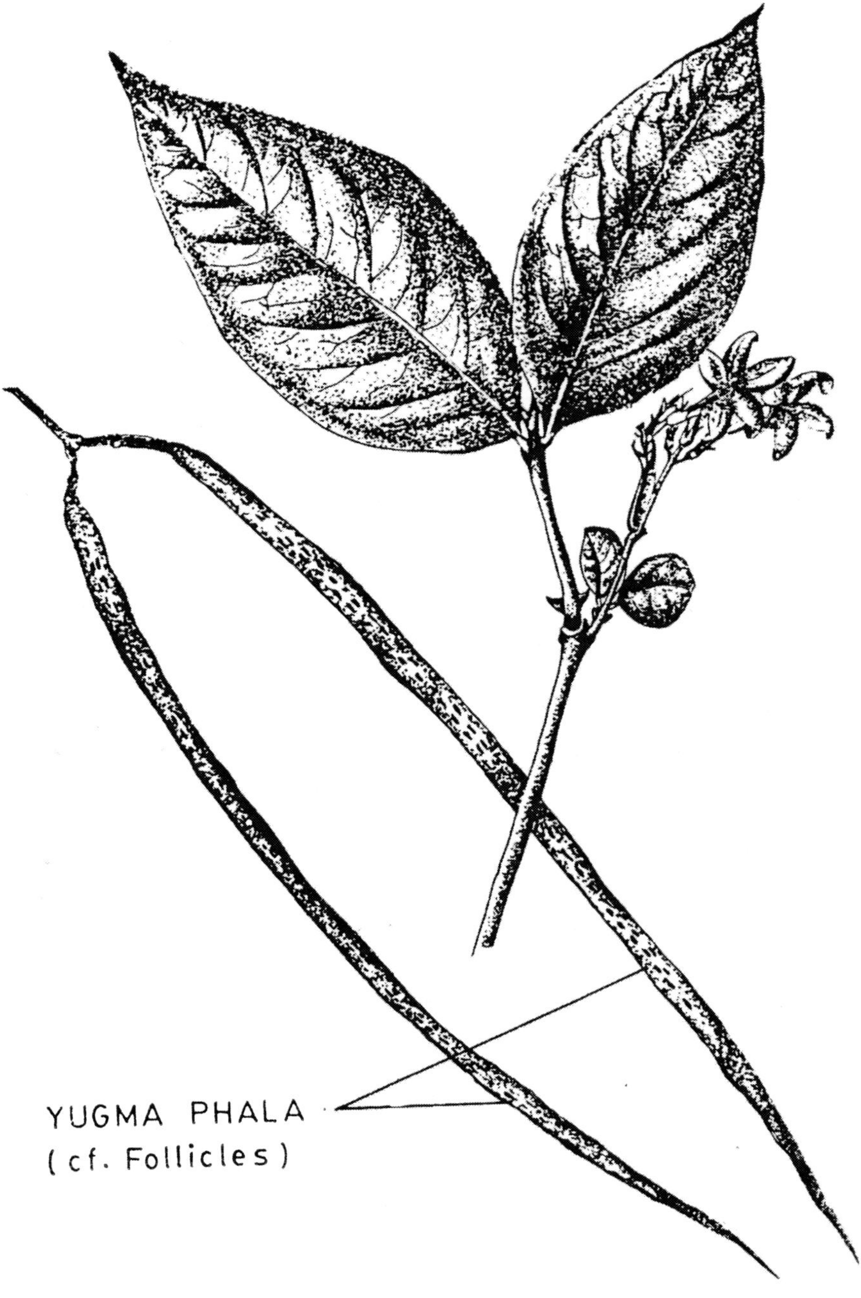

**Fig. 6 : (a) Kuṭaja (*Holarrhena antidysenterica = Ḥ pubescens*)
Mallikāpuṣpa Gaṇa (cf. Family Apocynaceae)**

Fig. 6 : (b) Saptaparṇa (*Alstonia scholaris* R.Br.)
Mallikāpuṣpa Gaṇa (cf. Family Apocynaceae)

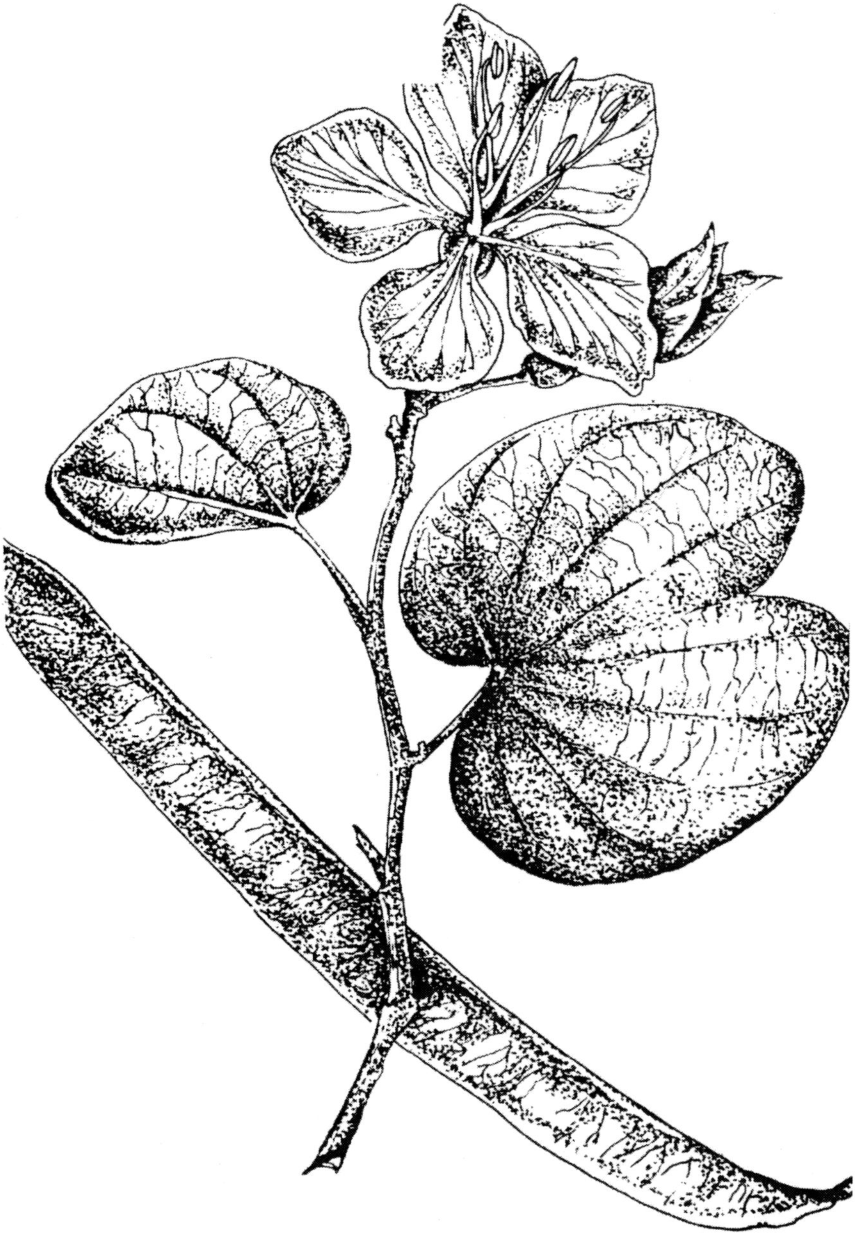

Fig. 6 : (c) Kovidāra (*Bauhinia purpuria* L.)
Vikarṇika (cf. Caesalpiniaceae)
Samī (cf. Family Leguminosae)

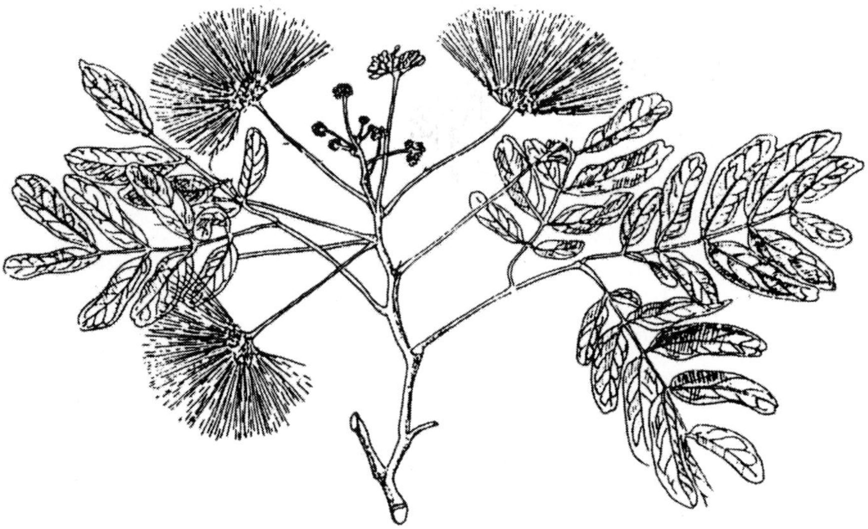

Fig. 6 : (d) Śirīṣa (*Albizzia lebbeck* Benth.)
Śūkapuṣpa (cf. Mimosaceae)
Samī (cf. Family Leguminosae)

Fig. 7 : A plant of 'Gulma kṣupa' order
Vāsaka (*Adhatoda vasica* Nees)
Simhapuṣpa (cf. Family Acanthaceae)

Fig. 8 :
(a) Dāru Haridrā (*Berberis aristata* DC.)
Kāṣṭha vallī

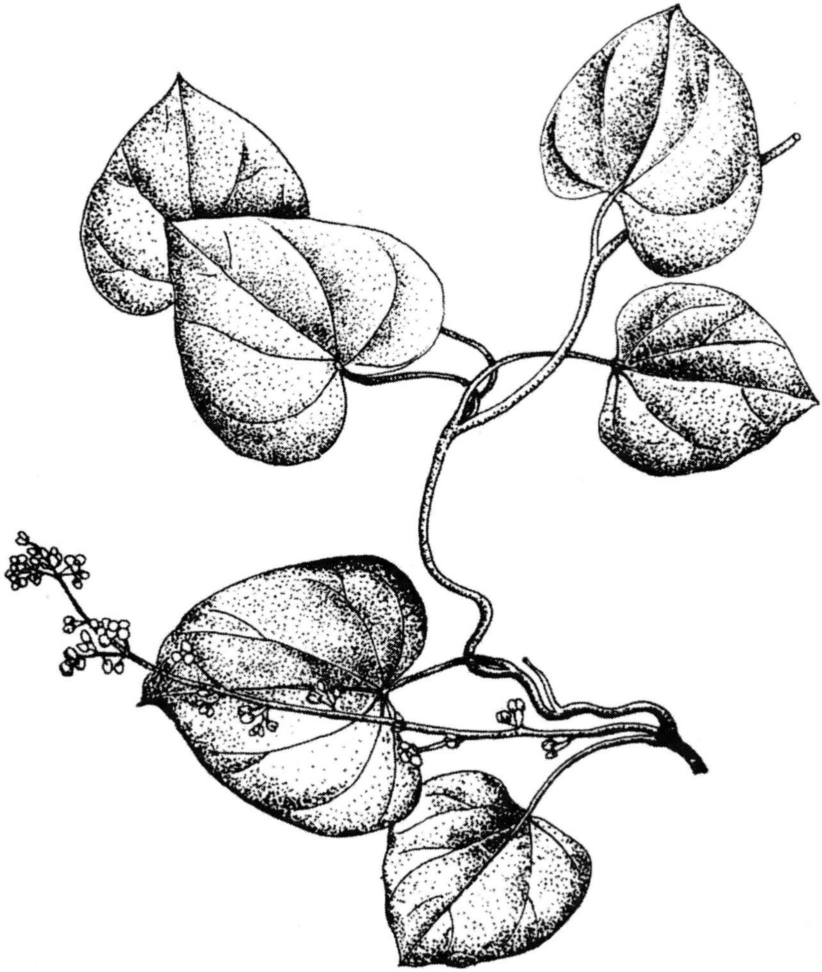

Fig. 8 :
(b) Amṛta vallī (*Tinospora cordifolia* Willd)

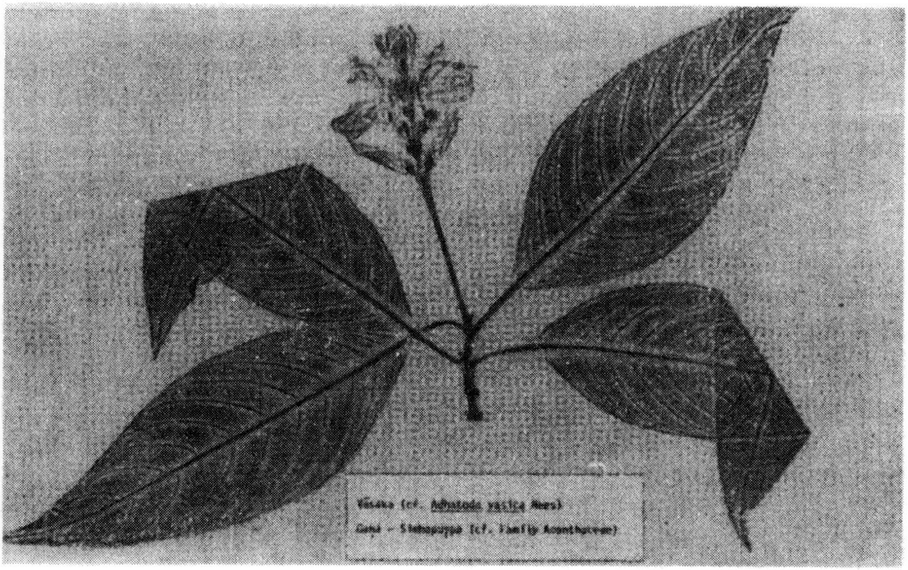

Fig. 9(a)

Brāhmi (Centella asiatica L.)
Gaṇa - Chatra (cf. Family Umbelliferae)

Fig. 9(b)

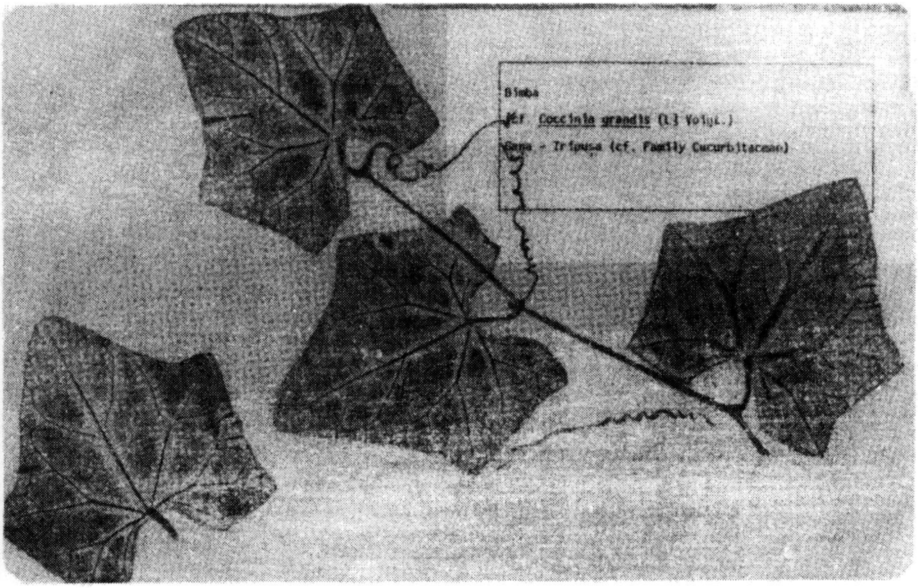

Fig. 9(c)

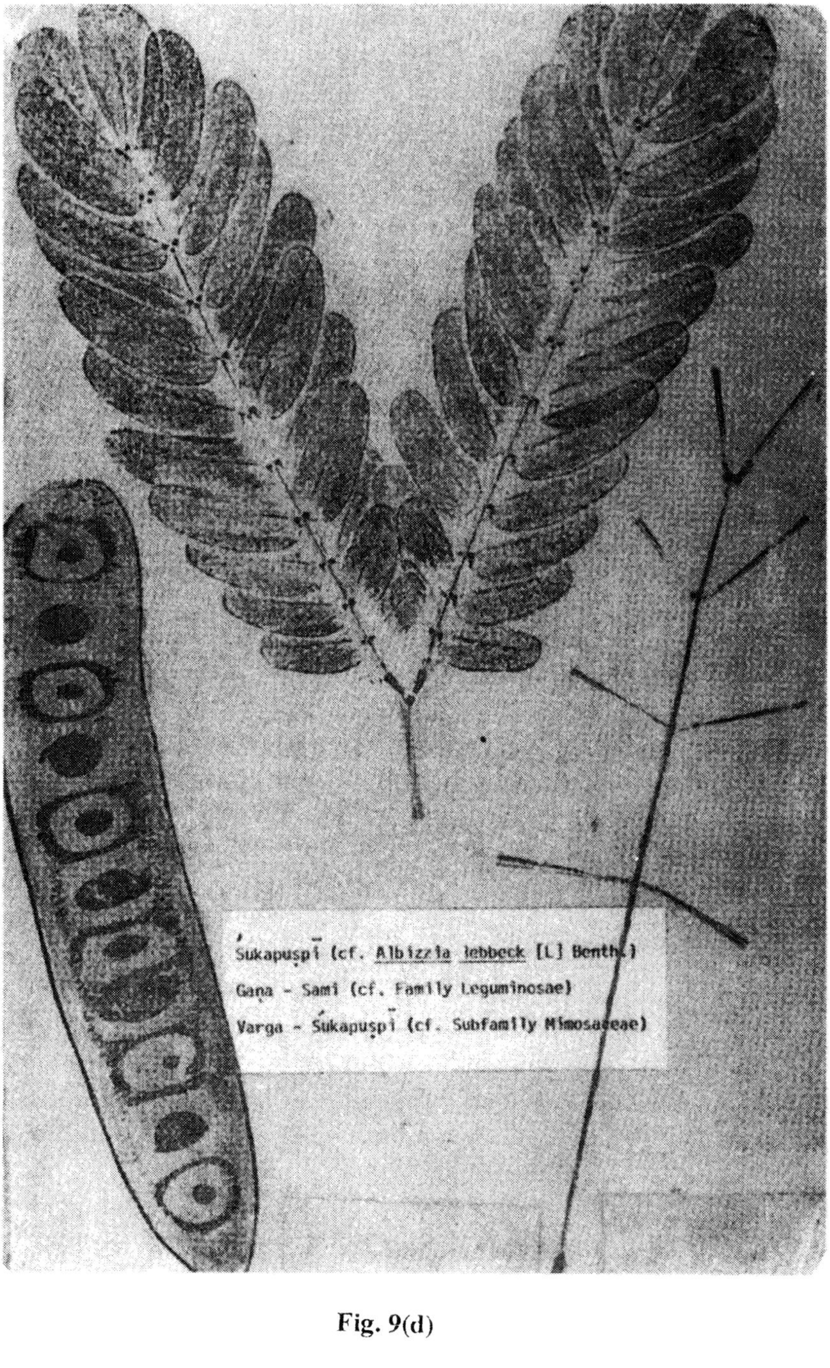
Śukapuṣpī (cf. Albizzia lebbeck [L] Benth.)
Gaṇa - Sami (cf. Family Leguminosae)
Varga - Śukapuṣpī (cf. Subfamily Mimosaceae)

Fig. 9(d)

Putikaranja
(cf. Caesalpinia bonduc [L] Roxb.)
gaṇa - Sami (cf. Family Leguminosae)
Varga - Vikarṇika (cf. Subfamily Caesalpiniaceae)

Fig. 9(e)

Fig. 9(f)